Introduction to
Data Analysis with R for Forensic Scientists

INTERNATIONAL FORENSIC SCIENCE AND INVESTIGATION SERIES

Series Editor: Max Houck

Firearms, the Law and Forensic Ballistics
T A Warlow
ISBN 9780748404322
1996

Scientific Examination of Documents: methods and techniques,
2nd edition
D Ellen
ISBN 9780748405800
1997

Forensic Investigation of Explosions
A Beveridge
ISBN 9780748405657
1998

Forensic Examination of Human Hair
J Robertson
ISBN 9780748405671
1999

Forensic Examination of Fibres,
2nd edition
J Robertson and M Grieve
ISBN 9780748408160
1999

Forensic Examination of Glass and Paint: analysis and interpretation
B Caddy
ISBN 9780748405794
2001

Forensic Speaker Identification
P Rose
ISBN 9780415271827
2002

Bitemark Evidence
B J Dorion
ISBN 9780824754143
2004

The Practice of Crime Scene Investigation
J Horswell
ISBN 9780748406098
2004

Fire Investigation
N Nic Daéid
ISBN 9780415248914
2004

Fingerprints and Other Ridge Skin Impressions
C Champod, C J Lennard, P Margot, and M Stoilovic
ISBN 9780415271752
2004

Firearms, the Law, and Forensic Ballistics, Second Edition
Tom Warlow
ISBN 9780415316019
2004

Forensic Computer Crime Investigation
Thomas A Johnson
ISBN 9780824724351
2005

Analytical and Practical Aspects of Drug Testing in Hair
Pascal Kintz
ISBN 9780849364501
2006

Nonhuman DNA Typing: theory and casework applications
Heather M Coyle
ISBN 9780824725938
2007

Chemical Analysis of Firearms, Ammunition, and Gunshot Residue
James Smyth Wallace
ISBN 9781420069662
2008

Forensic Science in Wildlife Investigations
Adrian Linacre
ISBN 9780849304101
2009

Scientific Method: applications in failure investigation and forensic science
Randall K. Noon
ISBN 9781420092806
2009

Forensic Epidemiology
Steven A. Koehler and Peggy A. Brown
ISBN 9781420063271
2009

Ethics and the Pracice of Forensic Science
Robin T Bowen
ISBN 9781420088939
2009

INTERNATIONAL FORENSIC SCIENCE
AND INVESTIGATION SERIES

Introduction to Data Analysis with R for Forensic Scientists

James Michael Curran

CRC Press
Taylor & Francis Group
Boca Raton London New York

CRC Press is an imprint of the
Taylor & Francis Group, an **informa** business

CRC Press
Taylor & Francis Group
6000 Broken Sound Parkway NW, Suite 300
Boca Raton, FL 33487-2742

First issued in paperback 2018

© 2011 by Taylor and Francis Group, LLC
CRC Press is an imprint of Taylor & Francis Group, an Informa business

No claim to original U.S. Government works

ISBN-13: 978-1-4200-8826-7 (hbk)
ISBN-13: 978-1-138-38144-5 (pbk)

This book contains information obtained from authentic and highly regarded sources. Reasonable efforts have made to publish reliable data and information, but the author and publisher cannot assume responsibility for the ity of all materials or the consequences of their use. The authors and publishers have attempted to trace the copy holders of all material reproduced in this publication and apologize to copyright holders if permission to publish i form has not been obtained. If any copyright material has not been acknowledged please write and let us know so w rectify in any future reprint.

Except as permitted under U.S. Copyright Law, no part of this book may be reprinted, reproduced, transmitted, c lized in any form by any electronic, mechanical, or other means, now known or hereafter invented, including photo ing, microfilming, and recording, or in any information storage or retrieval system, without written permission fro publishers.

For permission to photocopy or use material electronically from this work, please access www.copyright.com (www.copyright.com/) or contact the Copyright Clearance Center, Inc. (CCC), 222 Rosewood Drive, Danvers, MA 978-750-8400. CCC is a not-for-profit organization that provides licenses and registration for a variety of user organizations that have been granted a photocopy license by the CCC, a separate system of payment has been arra

Trademark Notice: Product or corporate names may be trademarks or registered trademarks, and are used o identification and explanation without intent to infringe.

Visit the Taylor & Francis Web site at
http://www.taylorandfrancis.com

and the CRC Press Web site at
http://www.crcpress.com

Dedication

This book is for Kevin Curran, Joseph Dickson, Vicky Gavigan, Bronagh Murray, Ken Tapper, Margie Tapper, Pat Tapper, Paul Tapper, Simon Tapper, and Libby Weston. Gone but not forgotten.

About the author

James M. Curran is currently an Associate Professor of Statistics in the Department of Statistics at the University of Auckland (Auckland, New Zealand). Dr. Curran is also the co-director of the New Zealand Bioinformatics Institute at the University of Auckland (www.bioinformatics.org.nz). His interests include computer programming, computational statistics, data analysis, forensic science, the design and analysis of experiments, and bioinformatics. Past research includes the interpretation and evaluation of forensic glass evidence, evaluation of forensic DNA evidence, and the effects of population substructure on DNA evidence.

Dr. Curran is the author of more than 70 scientific articles on the subject of statistical evaluation of forensic evidence. He has lectured statistics at the university level for 16 years and has been an invited speaker to many international workshops and conferences. He also also testified in forensic cases in the United States and Australia as an expert witness.

Dr. Curran is a Senior Consulting Forensic Scientist to the United Kingdom Forensic Science Service (FSS) where he has worked since 2001 with the development team for the FSS-i^3 software suite for DNA analysis. Dr. Curran is responsible for Kwikmix-3, a mixture interpretation package, and LoComatioN, which is an interpretation tool for low template DNA. Dr. Curran has also written the computational core of SPURS3, a package for DNA interpretation used in ESR, New Zealand.

When Dr. Curran is not at work he can be found on his road bike, in the garden, or playing with his PS/3, Xbox, or numerous computers.

Acknowledgments

There are many people to thank for their help and input to this book. Firstly, I would like to thank David Firth for hosting me at the University of Warwick where about two thirds of the manuscript was written. Thanks go to the University of Auckland and the UK Forensic Science Service for the financial support that made this work possible.

I would also like to thank those people who provided me with data: Rachel Bennett, John Buckleton, Sally Coulson, Tacha Hicks-Champod, David Lucy, Angus Newton, Ray Palmer, Jill Vintner, Simon Wong, and Grzegorz Zadora. Thanks also to José Almirall, Rod Gullberg, and Abderrahmane Merzouki who gave me data that I did not use in the end. I especially hope that all of the people who gave me data get something out of this book. Science is about transparency and repeatability. Access to data from experiments is vital in this respect, as it not only lets others see what you have done but also allows them to repeat your statistical analysis.

Thanks to those people who have helped with proofreading and general comments: Karin Curran, David Firth, Cedric Neumann, Kathy Ruggiero, and Torben Tvedebrink.

Thanks to those people who have helped me in my career over the years: Colin Aitken, David Balding, Martin Bill, John Buckleton, Christophe Champod, Ian Evett, Alan Lee, Roberto Puch-Solis, Alastair Scott, Chris Triggs, Kevan Walsh, and Simon Walsh.

And finally thanks to my family and my friends who put up with me.

Contents

1 Introduction   **1**
1.1 Who is this book for? 1
1.2 What this book is not about 1
1.3 How to read this book 2
    1.3.1 Examples and tutorials 3
1.4 How this book was written 4
1.5 Why R? 4
    1.5.1 R is free 4
    1.5.2 R does not have to be installed into system directories 5
    1.5.3 R is extensible 5
    1.5.4 R has a high-quality graphics system 5
    1.5.5 R allows you to share your analyses with others 6

2 Basic statistics   **7**
2.1 Who should read this chapter? 7
2.2 Introduction 7
2.3 Definitions 8
    2.3.1 Data sets, observations, and variables 8
    2.3.2 Types of variables 8
        2.3.2.1 Quantitative or qualitative 8
        2.3.2.2 Continuous, discrete, nominal, and ordinal . 9
2.4 Simple descriptive statistics 9
    2.4.1 Labeling the observations 10
    2.4.2 The sample mean, standard deviation, and variance . 10
    2.4.3 Order statistics, medians, quartiles, and quantiles . . . 12
2.5 Summarizing data 13
    2.5.1 An important question 13
    2.5.2 Univariate data analysis 14
    2.5.3 Three situations 14
    2.5.4 Two categorical variables 14
        2.5.4.1 Comparing two proportions 16
    2.5.5 Comparing groups 17
        2.5.5.1 Measures of location or center 17
        2.5.5.2 Measures of scale or spread 18
        2.5.5.3 Distributional shape and other features . . . 20
        2.5.5.4 Example 2.1—Comparing grouped data . . . 21

		2.5.6	Two quantitative variables	23
			2.5.6.1 Two quantitative variables—a case study . .	25
		2.5.7	Closing remarks for the chapter	27
	2.6	Installing R on your computer		27
	2.7	Reading data into R		28
		2.7.1	`read.csv`	29
			2.7.1.1 Checking your data has loaded correctly . . .	31
		2.7.2	`scan` and others	31
	2.8	The `dafs` package		32
	2.9	R tutorial		33
		2.9.1	Three simple things	33
			2.9.1.1 Tutorial	34
		2.9.2	R data types and manipulating R objects	38
			2.9.2.1 Tutorial	40
3	**Graphics**			**45**
	3.1	Who should read this chapter?		45
	3.2	Introduction		45
		3.2.1	A little bit of language	45
	3.3	Why are we doing this?		46
	3.4	Flexible versus "canned"		46
	3.5	Drawing simple graphs		46
		3.5.1	Basic plotting tools	47
			3.5.1.1 The bar plot	47
		3.5.2	The histogram	47
		3.5.3	Kernel density estimates	47
		3.5.4	Box plots	50
		3.5.5	Scatter plots	51
		3.5.6	Plotting categorical data	51
			3.5.6.1 Plotting groups	53
			3.5.6.2 Pie graphs, perspective, and other distractions	53
		3.5.7	One categorical and one continuous variable	56
			3.5.7.1 Comparing distributional shape	57
		3.5.8	Two quantitative variables	59
	3.6	Annotating and embellishing plots		60
		3.6.1	Legends	60
		3.6.2	Lines and smoothers	61
			3.6.2.1 Smoothers	61
		3.6.3	Text and point highlighting	62
		3.6.4	Color	63
		3.6.5	Arrows, circles, and everything else	63
	3.7	R graphics tutorial		64
		3.7.1	Drawing bar plots	64
		3.7.2	Drawing histograms and kernel density estimates . . .	68
		3.7.3	Drawing box plots	70

		3.7.4	Drawing scatter plots	71
		3.7.5	Getting your graph out of R and into another program	73
			3.7.5.1 Bitmap and vector graphic file formats	74
			3.7.5.2 Using R commands to save graphs	75
	3.8	Further reading		76
4	**Hypothesis tests and sampling theory**			**79**
	4.1	Who should read this chapter?		79
	4.2	Topics covered in this chapter		79
	4.3	Additional reading		80
	4.4	Statistical distributions		80
		4.4.1	Some concepts and notation	80
		4.4.2	The normal distribution	82
		4.4.3	Student's t-distribution	85
		4.4.4	The binomial distribution	86
		4.4.5	The Poisson distribution	87
		4.4.6	The χ^2-distribution	87
		4.4.7	The F-distribution	87
		4.4.8	Distribution terminology	87
	4.5	Introduction to statistical hypothesis testing		88
		4.5.1	Statistical inference	88
			4.5.1.1 Notation	88
		4.5.2	A general framework for hypothesis tests	89
		4.5.3	Confidence intervals	91
			4.5.3.1 The relationship between hypothesis tests and confidence intervals	92
		4.5.4	Statistically significant, significance level, significantly different, confidence, and other confusing phrases	93
		4.5.5	The two sample t-test	94
			4.5.5.1 Example 4.1—Differences in RI of different glass strata	94
			4.5.5.2 Example 4.2—Difference in RI between bulk and near-float surface glass	96
		4.5.6	The sampling distribution of the sample mean and other statistics	98
		4.5.7	The χ^2-test of independence	102
			4.5.7.1 Example 4.3—Occipital squamous bone widths 104	
			4.5.7.2 Comparing two proportions	105
			4.5.7.3 Example 4.4—Comparing two proportions relating to occipital squamous bones	106
			4.5.7.4 Example 4.5—SIDS and extramedullary haematopoiesis	107
			4.5.7.5 Fisher's exact test	107
			4.5.7.6 Example 4.6—Using Fisher's exact test	108

		4.5.7.7	Example 4.7—Age and gender of victims of crime	110
4.6	Tutorial			111

5 The linear model — 117

- 5.1 Who should read this? … 117
- 5.2 How to read this chapter … 117
- 5.3 Simple linear regression … 118
  - 5.3.1 Example 5.1—Manganese and barium … 119
  - 5.3.2 Example 5.2—DPD and age estimation … 121
    - 5.3.2.1 The normal Q-Q plot … 127
  - 5.3.3 Zero intercept models or regression through the origin … 128
  - 5.3.4 Tutorial … 129
- 5.4 Multiple linear regression … 133
  - 5.4.1 Example 5.3—Range of fire estimation … 133
  - 5.4.2 Example 5.4—Elemental concentration in beer bottles … 140
  - 5.4.3 Example 5.5—Age estimation from teeth … 142
  - 5.4.4 Example 5.6—Regression with derived variables … 146
  - 5.4.5 Tutorial … 146
- 5.5 Calibration in the simple linear regression case … 151
  - 5.5.1 Example 5.7—Calibration of RI measurements … 153
  - 5.5.2 Example 5.8—Calibration in range of fire experiments … 155
  - 5.5.3 Tutorial … 156
- 5.6 Regression with factors … 160
  - 5.6.1 Example 5.9—Dummy variables in regression … 163
  - 5.6.2 Example 5.10—Dummy variables in regression II … 164
  - 5.6.3 A pitfall for the unwary … 166
  - 5.6.4 Tutorial … 167
- 5.7 Linear models for grouped data—One-way ANOVA … 168
  - 5.7.1 Example 5.11—RI differences … 170
  - 5.7.2 Three procedures for multiple comparisons … 174
    - 5.7.2.1 Bonferroni's correction … 174
    - 5.7.2.2 Fisher's protected least significant difference (LSD) … 175
    - 5.7.2.3 Tukey's Honestly Significant Difference (HSD) or the Tukey-Kramer method … 176
    - 5.7.2.4 Which method? … 177
    - 5.7.2.5 Linear contrasts … 178
  - 5.7.3 Dropping the assumption of equal variances … 180
    - 5.7.3.1 Example 5.12—GHB concentration in urine … 181
    - 5.7.3.2 An alternative procedure for estimating the weights … 182
    - 5.7.3.3 Example 5.13—Weighted least squares … 183
  - 5.7.4 Tutorial … 183
- 5.8 Two-way ANOVA … 193

		5.8.1	The hypotheses for two-way ANOVA models	195
		5.8.2	Example 5.14—DNA left on drinking containers	196
		5.8.3	Tutorial	201
	5.9	Unifying the linear model		208
		5.9.1	The ANOVA identity	208

6 Modeling count and proportion data — 211
- 6.1 Who should read this? ... 211
- 6.2 How to read this chapter ... 211
- 6.3 Introduction to GLMs ... 212
- 6.4 Poisson regression or Poisson GLMs ... 213
  - 6.4.1 Example 6.1—Glass fragments on the ground ... 213
- 6.5 The negative binomial GLM ... 219
  - 6.5.1 Example 6.2—Over–dispersed data ... 220
  - 6.5.2 Example 6.3—Thoracic injuries in car crashes ... 223
  - 6.5.3 Example 6.4—Over-dispersion in car crash data ... 223
  - 6.5.4 Tutorial ... 225
- 6.6 Logistic regression or the binomial GLM ... 234
  - 6.6.1 Example 6.5—Logistic regression for SIDS risks ... 236
  - 6.6.2 Logistic regression with quantitative explanatory variables ... 237
  - 6.6.3 Example 6.6—Carbohydrate deficient transferrin as a predictor of alcohol abuse ... 237
  - 6.6.4 Example 6.7—Morphine concentration ratios as a predictor of acute morphine deaths ... 240
  - 6.6.5 Example 6.8—Risk factors for thoracic injuries ... 242
  - 6.6.6 Pitfalls for the unwary ... 243
  - 6.6.7 Example 6.9—Complete separation of the response in logistic regression ... 245
  - 6.6.8 Tutorial ... 245
- 6.7 Deviance ... 253

7 The design of experiments — 257
- 7.1 Introduction ... 257
- 7.2 Who should read this chapter? ... 258
- 7.3 What is an experiment? ... 258
- 7.4 The components of an experiment ... 259
  - 7.4.1 Questions of interest? ... 259
  - 7.4.2 Response variables ... 259
  - 7.4.3 Treatment factors ... 260
  - 7.4.4 Experimental units ... 260
  - 7.4.5 Structure in experimental units ... 261
  - 7.4.6 Assignment of treatments to experimental units ... 261
- 7.5 The principles of experimental design ... 261
  - 7.5.1 Replication ... 262

		7.5.2	Blocking	262
		7.5.3	Randomization	263
	7.6	The description and analysis of experiments		263
	7.7	Fixed and random effects		263
	7.8	Completely randomized designs		264
		7.8.1	Examples	264
			7.8.1.1 Block structure	265
			7.8.1.2 Treatment structure	265
			7.8.1.3 Randomization	265
			7.8.1.4 Analysis in R	266
			7.8.1.5 Factorial treatment structure	267
			7.8.1.6 Interaction plots	269
			7.8.1.7 Quantitative factors	269
	7.9	Randomized complete block designs		272
		7.9.1	Block structure	272
		7.9.2	Data model for RCBDs	272
			7.9.2.1 Example 7.1—Annealing of glass	272
			7.9.2.2 Treatment structure	273
			7.9.2.3 Block structure	273
			7.9.2.4 Tutorial: Analysis in R	273
			7.9.2.5 Example 7.2—DNA left on drinking containers	276
			7.9.2.6 Example 7.3—Blood alcohol determination	278
			7.9.2.7 Treatment structure	278
			7.9.2.8 Block structure	278
			7.9.2.9 Tutorial - analysis in R	278
		7.9.3	Randomized block designs and repeated measures experiments	281
			7.9.3.1 Example 7.4—Musket shot	282
			7.9.3.2 Treatment structure	284
			7.9.3.3 Blocking structure	284
			7.9.3.4 Tutorial - Analysis in R	284
	7.10	Designs with fewer experimental units		290
		7.10.1	Balanced incomplete block designs	290
			7.10.1.1 Example 7.5—DNA left on drinking containers II	291
		7.10.2	2^p factorial experiments	291
	7.11	Further reading		292

Bibliography 295

Index 301

Example Index 309

List of Figures

2.1	Descriptions of distributional shape	18
2.2	Refractive index (RI) of glass from New Zealand case work	19
2.3	Refractive index of float glass, including the float surface	19
2.4	Four data sets that have highly dependent variables but not necessarily with high correlation coefficients	24
2.5	The Anscombe quartet: all four data sets have the same correlation of 0.82	24
2.6	Correlation between two discrete variables	26
2.7	Pairs plot for bottle data	26
2.8	The R console	29
3.1	Constructing a kernel density estimate for 50 RI measurements from the same source	48
3.2	Histogram and KDE of all 490 RI values	49
3.3	Three typical kernels and their effect on the density estimates for the RI data	49
3.4	The components of a box and whisker plot	50
3.5	Bar plot of the genotype proportions for the Gc locus regardless of race code	51
3.6	Different colors of fibers found in human hair ($n = 12,149$)	52
3.7	Genotypes for the Gc locus by race code	53
3.8	Pie graph versus bar plot of Gc genotype proportions	54
3.9	Using pie graphs for comparison of groups is difficult	55
3.10	A stacked bar plot genotypes for the Gc locus by race code	56
3.11	Refractive index of a sample of float fragments by stratum	57
3.12	Kernel density estimates and histograms of the float glass RI by stratum	58
3.13	Zirconium concentration (Zr) in six different beer bottles	59
3.14	Concentration of manganese (Mn) versus barium (Ba) in six beer bottles	61
3.15	Fitting a locally weighted regression line or lowess line	62
3.16	The context menu from the R plotting window	73
3.17	A bitmap circle does not scale well	75
4.1	Normal distributions with the same standard deviation and different means	83

4.2	Normal distributions with the same mean but different standard deviations	84
4.3	Student's t-distribution gets a lower peak and fatter tails as the degrees of freedom decrease	85
4.4	A visual representation of a P-value	90
4.5	Box plot of RI measurements for the float surface (FS) and bulk (B) glass	95
4.6	Box plot of RI measurements for the near float surface (NFS) and bulk (B) glass	97
4.7	The probability function for the outcome of each experiment	99
4.8	The probability function for S, the sum of the two experiments	99
4.9	The Central Limit Theorem in action	100
4.10	Examining the two sample t-test for bulk glass and near float surface glass in more detail	102
4.11	Occipital squamous bone category frequency by race	104
4.12	Plot of RI for the 1st and 3rd panels from the Bennett data set	112
5.1	Scatter plot of age by deoxypyridinoline (DPD) ratio	122
5.2	Age versus DPD ratio with fitted line (dashed)	123
5.3	Residuaals versus predicted value (Pred-res) plots for three different regression situations	124
5.4	Pred-res plots magnify non-linearity. The quadratic trend here is magnified.	125
5.5	Pred-res plot for `age` versus `dpd.ratio` model	126
5.6	Possible shapes for a normal Q-Q plot. The theoretical quantiles are plotted on the x-axis and the empirical quantiles are plotted on the y-axis	127
5.7	A histogram of the residuals and a normal Q-Q plot of the residuals from the DPD fitted model	128
5.8	The `normcheck` function applied to the manganese and barium model	131
5.9	Pellet pattern area (in^2) versus firing range (ft)	134
5.10	Log area versus firing range (ft)	135
5.11	Pred-res plot for the shotgun scatter experiment	136
5.12	The shape of a quadratic	136
5.13	Diagnostic plots for the quadratic shotgun model	138
5.14	The dangers of over-parameterization. R^2 always increases with the addition of model terms.	139
5.15	Pairs plot for bottle data	140
5.16	Diagnostic plots for the manganese model	142
5.17	Pairs plot of the variables in Gustafson's teeth data	143
5.18	Regression diagnostic plots for teeth data	144
5.19	Prediction intervals for the Gustafson data using a multiple linear regression model and a simple linear regression model on a derived variable	147

5.20 The fitted quadratic model on the original scale for the shotgun training data 149
5.21 The classical calibration estimator and confidence interval .. 153
5.22 A calibration experiment for a GRIM2 instrument 154
5.23 A calibration line for a GRIM2 instrument........... 155
5.24 Predictions and prediction intervals for the shotgun experiment 157
5.25 A comparison of the prediction intervals produced by the classical method and the inverse regression method for the shotgun data. The dashed line represents the true range. 160
5.26 A different intercepts model for the shotgun experiment ... 163
5.27 Regression diagnostic plots for the RI calibration model ... 165
5.28 Different slopes model for GRIM2 calibration 166
5.29 A plot of three different panels from the Bennett data set .. 172
5.30 Residual diagnostic plots for a one-way ANOVA model 173
5.31 Gamma-hydroxybutyric (GHB) acid concentration in three groups of people................... 181
5.32 95% confidence intervals for the (log) GHB data using unweighted and weighted least squares 182
5.33 The (scaled) RI values plotted by group with observation numbers used as the plotting character 185
5.34 The probability that GHB measurement would fall below a threshold given that the person had received GHB 191
5.35 An example of interaction 194
5.36 Diagnostic plots for the two-way ANOVA model on the Abaz et al. data................... 198
5.37 Diagnostic plots for the two-way ANOVA model on the $\log_{10}$ Abaz et al. data 198

6.1 Plot of fitted versus observed values for the Wong data.... 216
6.2 Pred-res plot for the Poisson regression on Wong's data ... 217
6.3 The relationship between the mean and the variance for the Poisson model of the Wong data 217
6.4 Half-normal plot for the residual from the Poisson model... 218
6.5 A pairs plot for the Wong data 226
6.6 Diagnostic plots for linear models fitted to the Wong data .. 227
6.7 Half-normal plots for the residuals of the two Poisson GLM models for the Wong data 229
6.8 Half-normal plot for the residuals of the negative binomial GLM for the Wong data 231
6.9 The logistic function maps $z \in [-\infty, \infty]$ to $[0, 1]$ 235
6.10 Mean VH-CDT concentration by alcoholism status 238
6.11 The fitted probability curve for the VH-CDT data 239
6.12 Plot of the log-ratio of morphine concentration for acute (A) and random (R) cases 241
6.13 The behavior of the logistic function as group overlap decreases 244

6.14 Response (alc) with respect to Age and VH-CDT concentration 244
6.15 Probability of alcohol abuse with sensitivity to LOD analysis limits 251
6.16 The likelihood for π given our Bernoulli data 254

7.1 Interaction plot for the shotgun experiment 270
7.2 Interaction plot of salting experiment data 279
7.3 Interaction plot for salting data 282
7.4 Pellet pattern size with and without intermediate targets for the Jauhari et al. data 283
7.5 Scatter plot of pellet scatter pattern size by distance 285

List of Tables

2.1	Differences and squared differences for the foot size data . . .	11
2.2	Count table of race code by genotype for the Gc locus	15
2.3	Count table of genotype by race code for the Gc locus	15
2.4	Count table of genotype by race code for the Gc locus	16
2.5	Frequency table of genotype by race code for the Gc locus . .	16
2.6	RI of 10 recovered glass fragments	20
2.7	$IQR(\alpha)$ for the recovered RI data and various values of α . .	21
2.8	The variables in the bottle data set	22
2.9	Summary statistics for Zr concentration by body part	22
3.1	Genotype proportions for the Gc locus regardless of race code	52
3.2	Number of misleading signatures in comparisons of 16 genuine signatures and 64 simulated signatures from 15 document examiners	60
3.3	R commands for various plotting devices/formats	76
4.1	Outcomes for the sum	100
4.2	Occipital squamous bone classifications by race	103
4.3	Proportion of EMH diagnoses for non-SIDS ($n = 102$) and SIDS ($n = 51$) cases	107
4.4	The expected cell counts for the EMH data under the assumption of independence	108
4.5	Cross classification of victims of crime by gender and age . .	110
5.1	Variables in Gustafson's teeth data [1]	142
5.2	Simple linear regression of Age against TP (total points) using the Gustafson data	146
5.3	Predicted firing range (ft) from the shotgun experiment . . .	156
5.4	Regression table for the shotgun data with a different intercepts model	163
5.5	ANOVA table for the RI data	172
5.6	Regression table for the RI one-way ANOVA model	174
5.7	Results of (two-tailed) significance tests for a difference between the two means. A + means that the test was significant at the 0.05 level or that the 95% confidence interval did not contain zero	178

5.8 Comparison of Welch-James and residualized weights 183
5.9 The different beverage and drinking container combinations in Abaz et al. 196
5.10 ANOVA table for Abaz et al. data 197
5.11 ANOVA table for the additive model for the Abaz et al. data 199
5.12 The mean for a linear model by technique 208
5.13 ANOVA table for the full multiple regression model on Gustafson's tooth data 209
5.14 Model terms as a percentage of (cumulative) variance explained 210

6.1 Analysis of deviance table for Poisson regression on Wong data 215
6.2 Analysis of deviance table for Poisson regression on Wong data with the variables re-ordered 215
6.3 Analysis of deviance table for Wong data with outliers removed 219
6.4 A fully saturated Poisson model for the Wong data 220
6.5 Analysis of deviance table for the Wong data using a quasi-Poisson model 221
6.6 Analysis of deviance table for the fully saturated Poisson GLM fitted to the Kent and Patrie data 224
6.7 Analysis of deviance table for the fully saturated quasi-Poisson GLM fitted to the Kent and Patrie data 224
6.8 Analysis of deviance table for the fully saturated negative binomial GLM fitted to the Kent and Patrie data 225
6.9 Summary table for the liver binomial GLM 236
6.10 Summary table for logistic regression of alcoholism status on VH-CDT concentration 238
6.11 A hypothetical confusion matrix 240
6.12 Confusion matrix for the logistic model 240
6.13 Regression summary for the morphine concentration data . . 241
6.14 Cross classification of the experiments by `gender`, `injury`, and `load.cond` 242
6.15 A logistic regression model for injury level in the crash data . 243

Chapter 1

Introduction

1.1 Who is this book for?

The title of this book is *Introduction to data analysis with R for forensic scientists*. I wrote it primarily with a forensic audience in mind, but really it is for anyone who is working in a laboratory. Basic experimental science generates a lot of data. The field of statistics provides a logical, coherent framework in which those data can be analyzed. This means that, if someone else takes our data and applies the same techniques, they should be able to at least get the same results and hopefully arrive at the same inferences and conclusions. In my reading of forensic science journals over the years, I have seen a lot of interesting research-generated data that has not been explored to its full potential. I theorized that this is because the researchers involved either did not know enough statistics or felt that they did not have the skills or software that would let them analyze the data in the appropriate way. This book is for them.

My aim in this book is to minimize the theory, minimize the mathematics, and concentrate on the application and practice of statistics. This does not mean that there is no theory or no mathematics in this book. What it means is that I have made a conscientious effort not to introduce theory or mathematics unless it is vital to the explanation of the problem. The approach, therefore, unlike many textbooks, is to introduce the relevant information when it is needed rather than to front-load a chapter with theory before we have seen any data.

1.2 What this book is not about

The focus of this book is not the statistical interpretation of forensic evidence. There are plenty of excellent books that cover that subject, such as Buckleton, Triggs, and Walsh [2] and Evett and Weir [3] for DNA, and Aitken and Taroni [4] for much of the rest. A lot of basic science has to be done before any substance or item can be used as evidence. This book is for people doing that science.

Neither is the book about Bayesian statistics or Bayesian evangelism. It is possible to do much of the data analysis described in this book in a Bayesian way. However, it is usually harder and requires a great deal of technical skill. Furthermore, although the interpretations are different, the numerical results are often the same or very similar. A Bayesian version of this book may come in the future, but for the time being, I have used traditional frequentist techniques. I have also concentrated on the understanding of the output. It is all very well being Bayesian, but if you do not understand the frequentist result, then you will not see the advantage.

Finally, I have not written anything about multivariate statistics. The application of multivariate statistical techniques to forensic data is an important and much-needed subject. However, it is my firm belief that to be good at multivariate statistics you need to have a solid understanding of univariate statistics. I would rather the reader concentrate on understanding the issues in this book rather than leaping into multivariate space where everything is harder to see and understand.

1.3 How to read this book

The book is divided up into seven chapters, including this one. Each of the chapters covers a major topic. These topics are

- **Chapter 2:** Basic statistics—tools for elementary numerical summary of data;

- **Chapter 3:** Graphics—considerations and techniques for the visual display of data;

- **Chapter 4:** Hypothesis tests and sampling theory—an overview of statistical hypothesis tests and the reasoning behind them;

- **Chapter 5:** The linear model—covers the major modeling tools of statistics;

- **Chapter 6:** Modeling count and proportion data—an introduction to extensions to the linear model for commonly encountered scenarios;

- **Chapter 7:** The design of experiments— an introduction to the (statistical) design of experiments

Chapters 2 and 4 can probably be regarded as a refresher for those readers who have done at least one course in statistics. Chapter 2, however, does provide

Introduction

an introduction to the use of R, which is important in all chapters. Chapter 3 is essential reading. Good-quality statistical graphics are almost more important than the statistical analysis. A well-designed graph can show the most important features of the data better than any statistical test. Chapter 5 is a comprehensive review of the linear model and its applications. The linear model is the foundation of most of the statistics you will have encountered to date. Chapter 6 is a specialty chapter for readers with problems involving count data or proportion data. The chapter covers logistic and Poisson regression, which are techniques not generally covered in a first course in statistics. Both techniques are extremely useful. Logistic regression in particular has application in the classification of objects to sources—one of the fundaments of forensic science. Chapter 7 should be read by anyone planning to carry out an experiment. Even if you do not do any of the examples, you will find that there are ideas there that will help you plan and design your next experiment so that you can minimize the cost of carrying out the experiment and maximize your chance of finding differences that may exist.

1.3.1 Examples and tutorials

The examples and tutorials in this book follow a simple pattern. The tutorial exercises repeat the analyses shown in the examples. The reason for this is that many people learn by copying the examples. Once you have a feel for what an example is showing you, then you can begin to apply the techniques to your own problem. The worst thing in a book like this for me is to be unable to repeat the examples. This is doubly frustrating when a computer program is involved. Therefore, the tutorials (for the most part) are step-by-step guides through the analyses shown in the examples. This structure also lets me give an example without cluttering the main text with computer code. The code is essential, and, of course, many readers would like to see it. This is why we work through it in the tutorials.

I urge all readers to work through as many of the tutorials as possible. It is only by doing these analyses yourself that you will truly get a feel for the practice of data analysis. You will also find that once you gain confidence that you will be able to use R to try out ideas of your own. Data analysis is as much an art as it is a science. There is no "one true way" to analyze a data set, and by "having a play," you will be able to try out your own analyses.

The data in all of the examples in this book are real. That means I have collected real data from real experiments. This was done through using data that I have obtained from experiments of my own, my students, and my colleagues, as well as published data from journals. The data for all of the examples are available in a free R package for this book. The details of this package and how to install it are given on page 32.

1.4 How this book was written

This book was written with a literate programming environment called Sweave [5]. Sweave allows the author to *weave* code into a document. A document preprocessor then executes the code and embeds the formatted code and output back into the document. R provides the preprocessing and LaTeX does the beautiful typesetting you see before you. What this means for you, the reader, is that **there are no syntax errors in this book**. It means that every piece of code you see in this book has been executed as written. Therefore, if you copy the code verbatim, you should get exactly the same results as I have. This, of course, does not prevent me from doing something foolish. However, it should have minimized the chances of readers encountering the ever-frustrating syntax error. As a child learning to program, I relied heavily on code listings in programming magazines. I would faithfully copy the BASIC code and occasionally pages and pages of hexadecimal into my computer. All too often, though, these programs had mistakes that rendered them unusable. I hope that will not happen with the code in this book. If, through some fault of mine, you find a mistake, please email me so that I may keep an up-to-date list of errata.

1.5 Why R?

I expect one of the most frequent questions that will be asked or comments that will be made about this book is "Why did you choose to use R and not something easy like Excel?" This is a common refrain from my students. The reasons are manyfold; however, I will attempt to give them all.

1. R is free
2. R does not have to be installed in system directories
3. R is the choice of many professional statisticians
4. R is extensible
5. R has a high-quality graphics system
6. R allows you to share your analyses with others

1.5.1 R is free

The first and foremost reason for choosing R is that R is completely and utterly free. It has been my experience that laboratory directors will spend hundreds of thousands of dollars on analytical machines but are reluctant to spend even a thousand dollars on software. R's price tag of $0 overcomes that obstacle immediately. An added bonus to R's being free is that users can always have the most up-to-date version simply by visiting http://r-project.org and downloading it from the nearest CRAN mirror.

1.5.2 R does not have to be installed into system directories

How many of you work on networked computers where you are not allowed to add or remove programs? R has a "Unix-style" installation structure. That means that all the files it needs are contained in a single directory (with structure below) and that it does not attempt to install files in system directories (like C:\Program Files under Microsoft Windows \usr\bin under Linux). This means you can install R in your own directory (where, presumably, you are allowed to save files) or even on a USB flash drive. R has no impact on the operating system, or at least no impact that affects its use.

1.5.3 R is extensible

One of the things I dislike most about programs such as Microsoft Excel is that it is very hard to extend or alter their "canned" routines. I am fully capable of writing Visual Basic or the like to overcome some of these hurdles, but many users are not. R is a statistical programming language. That means that although it comes with many programs or functions to perform statistical analyses, the user is not constrained to using solely those programs. The user can write their own programs or include the programs of others to help them do the task that they want to do, not the task that the program thinks they should do. It is this extensibility that makes R the success that it is because anyone who has a specialist research area can write, and generally find, programs that help them solve the problems they encounter in this area.

1.5.4 R has a high-quality graphics system

R has a high-quality highly flexible graphics system that can produce production-quality graphics. Unlike other programs, you are not constrained to the default "chart-types" that the system can produce. R has many built-in graphing functions, and, in their simplest form, most will produce a very good graph with a single command. However, should you, for example, want to add a legend, color the points, add a smoothing line, or draw a box plot on top of a scatter plot, then you can do it. R regards a graph as a blank canvas on which you can draw. Furthermore, considerable thought has gone into presenting data so that the information contained therein is not obscured by chart junk such as grid lines on a graph, colored backgrounds, lines joining points when there is no reason to connect the points, and meaningless legends such as "Series 1."

1.5.5 R allows you to share your analyses with others

One task I find myself doing quite often is repeating a statistical analysis that someone has done in Excel. It can be quite a chore figuring out which variables have been used and where those variables live within the worksheet or workbook. An R program (or script) makes all the steps in an analysis quite explicit, because at the end of the day, it is a program—a sequential set of instructions that allows the user to tell the computer what to do. This means that if someone else has your data files and your R script, then they should be able to easily replicate your analysis. Replication of results is an important part of any scientific discipline.

Chapter 2

Basic statistics

A statistical analysis, properly conducted, is a delicate dissection of uncertainties, a surgery of suppositions.—M. J. Moroney, Statistician.

2.1 Who should read this chapter?

This chapter provides a revision of simple summary statistics and an introduction to R. If you have done a freshman course in statistics, then I suggest you skim this chapter and work through the tutorial at the end. If you have never done any statistics, then I would suggest you read this chapter thoroughly. We will rely heavily on an understanding of means and variances later on and of course on the R skills that you will start to learn in this chapter. If you are skilled in both statistics and R, then you can skip this chapter altogether. I would suggest, however, that you skim the section on summarizing data. Good tabular display of data is as important as a good graph.

2.2 Introduction

In this chapter, we will introduce some of the basic statistical concepts. In particular, we cover the categorization of variables, simple descriptive statistics, and the numerical summary of data. This chapter also provides instruction on how to install R and a first tutorial in R.

2.3 Definitions

It is useful to know how a statistician thinks about data. It is useful because I am a statistician, and I am going to use various words to describe things that I am familiar with.

2.3.1 Data sets, observations, and variables

Most statisticians think of a **data set** in the same way that you might think of a spreadsheet. The data are laid out in a series of rows and columns. Each row contains an **observation**, which is a set of one or more pieces of information about a single instance, individual, or object. The pieces of information recorded for each observation are called **variables**. For example, you might imagine a medical situation where your observations consist of information about patients. In this situation, you might record the patient's name, age, gender, medical record number, and perhaps whether they are alive or dead. In this example, each patient is an observation, and *name*, *age*, *gender*, *record number*, and *status* are our variables. Returning to our spreadsheet analogy, the rows of the sheet are the observations, the columns of the sheet are the variables, and the whole sheet is the data set.

2.3.2 Types of variables

It is useful to have a set of terms that describe what type of information a variable might record. These terms can be used later on to guide how we might present or analyze data. They also can be used to guide our questions about the data and about the experiment, survey, or process that produced the data.

2.3.2.1 Quantitative or qualitative

The description of variables is hierarchical. At the first level, we divide variables into being **quantitative** or **qualitative**. A quantitative variable is, at its simplest, one that records a number. Some examples of quantitative variables are length in millimeters, time in seconds, age in years, and the number of occurrences of an event. A qualitative variable, on the other hand, records a description or a category. For example, car color is a qualitative variable. We might record car color as *blue*, *red*, *white*, or *other*. Alternative terms that statisticians use for a qualitative variable are **categorical** and **dichotomous** or **polychotomous**. I use *categorical* frequently. The latter two are used in the scientific and social science literature and are mentioned for completeness rather than clarity! I say this because many people incorrectly use "dichotomous" when there are more than two categories.

2.3.2.2 Continuous, discrete, nominal, and ordinal

The next level of the hierarchy subdivides the quantitative and qualitative descriptions.

We can describe a quantitative variable as being either **continuous** or **discrete**. When using these terms, it is usual to drop the word "quantitative." That is, we talk about continuous or discrete variables with the implicit assumption that both of these are quantitative variables. One way to think about continuous variables is that there are "no gaps" between the possible values. In theory, this means that no matter how fine the scale on which we measure something, we can always measure it more finely. Take length as an example. We can measure something in meters, in centimeters, in millimeters, or in micrometers (10^{-6} mm). In practice we know there are limits, but the concept remains. A discrete variable has gaps between the possible outcomes. Discrete variables most often occur when we are counting something, e.g., the number of Caucasians using a particular bus route in an hour, or the number of pieces of glass found on a suspect. These definitions mostly work, but there are circumstances where we may treat what is strictly a discrete variable as continuous and vice versa.

We can describe a qualitative variable as being either **nominal** or **ordinal**. It is easier to describe ordinal variables first. An ordinal variable, as the name suggests, has a natural ordering in the categories. For example, we might record a temperature as *low, medium,* or *high*. Another example comes from the US Centers for Disease Control and Prevention (CDC), which classifies diseases in levels of bio-hazard ranging from Level 1 (minimum risk) to Level 4 (extreme risk). A nominal variable has no natural ordering. For example, I cannot conceive of a natural ordering for car colors.[1] Most statisticians usually do not make the distinction between nominal and ordinal variables and simply refer to them as categorical. The reasons that these distinctions exist is that there are techniques that make use of ordinal information. In addition, there are some comparisons that make more sense with ordinal data. For example, we are likely to be interested in the differences between groups of observations in the high category versus groups of observations in the low category, whereas there would be fewer reasons why a particular comparison between different colors of cars would be sensible.

2.4 Simple descriptive statistics

Most people are familiar with some descriptive statistics such as the average or mean, the median, and perhaps the standard deviation. In this section,

---

[1] Please do not email me telling me about RGB or other color classification systems!

I will take the opportunity to refresh your memory, and to introduce some statistics that perhaps you have not used before. I also want to spend some time explaining how these statistics work and when they might be useful. I will attempt to keep the formulas to a minimum as I promised in the start of this book.

2.4.1 Labeling the observations

We need a tiny bit of notation to make any sense of the ensuing material. In this section, I will often talk about a generic variable, X, rather than a specific example on which we have made a set of n observations. We label the observations $x_1, x_2, \ldots, x_n$. Just as an example, imagine that X is foot size (in mm), and we have measured the foot size of 5 males. The observations, to the nearest 0.1 mm, are 271.7, 261.6, 243, 236.7, and 259.5. That is, $x_1 = 271.7$, $x_2 = 261.6, \ldots,$ $x_5 = 259.5$. Why X and n rather than some other letters or symbols? There is no particular reason to choose X and n, but there are some useful conventions in statistical notation. We use uppercase letters to represent random variables and lowercase letters to represent either observations on a variable or things we regard as constant (like n the sample size).

2.4.2 The sample mean, standard deviation, and variance

Most scientists are familiar with the average or mean and standard deviation of a set of observations. Statisticians talk about the **sample mean** and the **sample standard deviation**. They use the word *sample* to say "this statistic comes from a specific sample."

We define the **sample mean** for a set of n observations on a variable X as

$$\bar{x} = \frac{1}{n} \sum_{i=1}^{n} x_i \qquad (2.1)$$

Equations like (2.1) mean nothing to many people. Sometimes it helps to see what the notation means in words. Equation (2.1) says

> Add all the observations $x_1, x_2, \ldots, x_n$ together and divide by the sample size n.

We define the **sample standard deviation** for a set of n observations on a variable X as

$$s = \sqrt{\frac{1}{n-1} \sum_{i=1}^{n} (x_i - \bar{x})^2} \qquad (2.2)$$

Equation (2.2) in words says

Basic statistics

> Add up all the squared differences between the observations $x_1, x_2, \ldots, x_n$ and the sample mean $\bar{x}$ and divide by the sample size minus 1, $n-1$, then take the square root.

It is probably useful to look at this formula in parts. I have found that my students tend to think of the standard deviation as a button on the calculator, or a menu item in Minitab or SPSS, and consequently do not have any feel for what the standard deviation measures. The standard deviation is a statistic that tells us about "the average distance of the observations from the mean." It is useful because it tells us how close or how spread out the data are relative to the mean. If the standard deviation is small relative to the mean, then the observations are tightly scattered around the mean. If the standard deviation is large relative to the mean, then the observations are widely spread out. You might notice that my definition does not quite match up with the formula. In fact, if I just took the words "the average distance of the observations from the mean," the formula I would write down is

$$\frac{1}{n}\sum_{i=1}^{n}(x_i - \bar{x}) \qquad (2.3)$$

The trouble with Equation (2.3) is that it always works out to be zero. This is simply a function of how the formula for the sample mean works. The (mathematical) solution to this problem is to use squared distances instead. It is easiest to see how this works with an example. Let us look at our foot size data again. The mean of our sample is

$$\bar{x} = \frac{1}{5}(271.7 + 261.6 + \cdots + 259.5)$$
$$= 254.5$$

I hope Table 2.1 will convince you that Equation (2.3) works out to be zero.

Obs. No.	Obs.	$x_i - \bar{x}$	$(x_i - \bar{x})^2$
1	271.7	17.2	295.84
2	261.6	7.1	50.41
3	243.0	-11.5	132.25
4	236.7	-17.8	316.84
5	259.5	5	25
Sum	1272.5	0.0	820.34

TABLE 2.1: Differences and squared differences for the foot size data

If it does not convince you, try it for yourself in Excel or on a calculator. We can see in Table 2.1 that the squared difference is always positive. This reflects our interest in the spread rather than the direction of the difference. That is, we do not really care whether an observation is above or below the mean, just how far away it is. If we were interested in the average squared difference, then all we would need to do is divide 820.34 by $n = 5$. However, Equation (2.2) divides the sum by $n-1$ and not n. The reason is technical and not important here. What is important to note is that dividing by n or $n-1$ makes almost no difference for large sample sizes. So let us assume we have (approximately) divided by the sample size. The last problem we are left with is that we have the average squared difference, rather than the average difference. There are two solutions here. We could have used the absolute value of the differences. That is, if the difference was negative we make it positive. This is a legitimate statistic and is called the **mean absolute deviation**, but it is not commonly used. The alternative is to take the square root to transform the differences back to the original scale. This is not strictly true because in general

$$\sqrt{a^2 + b^2} \neq a + b$$

For example,

$$\sqrt{1^2 + 2^2} = \sqrt{1 + 4}$$
$$= \sqrt{5} \neq 3$$

However, it provides us with a useful way of thinking about it.

Statisticians quite often talk about the **sample variance** instead of the sample standard deviation. The sample variance is usually denoted by s^2 and is simply the sample standard deviation squared. The sample standard deviation of our example data is $s = 14.32$, and the sample variance is $s^2 = 14.32^2 = 205.085$.

2.4.3 Order statistics, medians, quartiles, and quantiles

The order statistics for a set of observations $x_1, x_2, \ldots, x_n$ are simply the observations sorted into ascending order. We denote the order statistics $x_{(1)}, x_{(2)}, \ldots, x_{(n)}$ and they obey the relation $x_{(1)} \leq x_{(2)} \leq \ldots \leq x_{(n)}$. For our example data, the order statistics are 236.7, 243, 259.5, 261.6, and 271.7. It should be obvious that for any data set $x_{(1)}$ is the **minimum**, and $x_{(n)}$ is the **maximum**. The order statistics let us define the **sample median**. The median is the "middle order statistic." That is,

$$median = \begin{cases} x_{\left(\frac{n}{2}\right)} & \text{if } n \text{ is odd,} \\ \frac{x_{\left(\frac{n}{2}-0.5\right)} + x_{\left(\frac{n}{2}+0.5\right)}}{2}, & \text{if } n \text{ is even} \end{cases} \quad (2.4)$$

We will leave the formulas behind at this point. The median and the lower and upper quartiles, which we have yet to define, are part of a wider set of statistics called **quantiles** or **percentiles**. The median is the 0.5 quantile or the 50th percentile. This means that 50% of the data set is less than or equal to the median (and 50% is greater than the median). The **lower** and **upper quartiles** are the 0.25 and 0.75 quantiles (or the 25th and 75th percentiles), respectively. In general, the α-quantile q_α is the value that $100\times\alpha\%$ of the data are less than. My students have typically found this definition very confusing. However, if we return to the median and the percentiles, then we can relate the definition to something we are familiar with. If one of our children takes a test and scores in the 90th percentile, then most of us instinctively know that this is good. It means that approximately 90% of the students who sat the test had a score that was equal or lower to our child's score. The 90th percentile is the 0.9 quantile, or $q_{0.9}$, and $100\times 0.9 = 90\%$ of the scores are less than $q_{0.9}$. Quantiles are useful in a number of situations. The median, or the 0.5 quantile, is a robust statistic for describing the center of a data set. The upper and lower quartiles and the distance between them, which is called the **interquartile range** or **midspread**, give us an indication of the spread of the data by telling us about the central 50% of the data. In other situations, such as the construction of confidence intervals, we, perhaps unknowingly, use the 0.025 and 0.975 quantiles to tell us about the central 95% of the data. We will also use the **sample quantiles** or **empirical quantiles** as a diagnostic tool for linear models.

2.5 Summarizing data

One of the simplest things we can do with a data set is provide a numerical summary. This usually involves calculating some of the descriptive statistics we have just seen for the quantitative variables or perhaps presenting frequency tables for categorical data.

2.5.1 An important question

We should always ask ourselves when summarizing numerical data

> "What am I hoping to show by providing this summary?"

Unfortunately, this question is not often asked. All too often data are summarized by inappropriate or meaningless statistics. After all, "The average human has one breast and one testicle" to quote Des McHale (Professor of Mathematics, University College Cork).

2.5.2 Univariate data analysis

Many introductory statistics textbooks start with the analysis of a single sample of data on a single variable. This situation is contrived. We almost never collect data on a single variable. Our aim in forensic science, as it is in many other fields of science, is to make comparisons and find and describe relationships. We might have information about a single quantitative feature of a sample of objects, but we will usually have information that divides our observations into groups. Alternatively, we might have measured two variables and want to know whether there is a relationship between them, e.g., is the concentration of one element in a substance related to the concentration of another. Therefore, by definition, we have two variables. For that reason, I am not going to spend any time discussing techniques for single variables. We are interested in making comparisons and finding relationships, and we are going to start there.

2.5.3 Three situations

I will assume for the time being that we have just two variables in our data set. Therefore, there are three possible combinations of variable types:

1. Two categorical variables
2. One categorical and one quantitative variable
3. Two continuous variables

These combinations guide us in our choice of summary. I will not discuss the analyses in this chapter of the book because each topic needs a chapter to itself. However, I will briefly cover the summaries.

2.5.4 Two categorical variables

If we have two categorical variables, then these variables divide our sample up into smaller groups that are formed by the combinations of the categories. This process is called **cross-classification** or **cross-tabulation**. For example, we might cross-classify a sample of people by their race code in a DNA database and their genotype at the human group-specific component (Gc) locus from the Perkin-Elmer Amplitype Polymarker DQ-α Typing Kit. The FBI has published a reference database that has the Gc genotypes of African Americans, Caucasians, and South Western Hispanics. The Gc locus has alleles A, B, and C and hence, the possible genotype are AA, AB, AC, BB, BC, and CC. Therefore, cross-classifying this data on race code and Gc genotype divides our sample into six groups. The aim of such a cross-classification can be looked at in two ways. We might ask whether the genotype proportions in the genotype categories are the same for each race code. Or, we might ask whether there is an association or dependence between Gc genotype and race code. These are equivalent questions, both statistically and logically. The best summary of this type of data is a table. Statistician Andrew Ehrenberg gave a set of six simple rules for presenting tabular data effectively [6]:

1. Round drastically – "two busy digits"
2. Arrange the numbers so that comparisons are made column-wise and not row-wise
3. Order the columns by size or use ordinal information
4. Use row and column averages or sums as a focus
5. Use white space well
6. Provide verbal summaries

We shall show how to apply these rules to the Gc data. Tables 2.2 and 2.3 show the counts of the observations falling into each of the race code and genotype categories. Given that our focus is comparing genotype proportions between categories, we invoke rule 2 and choose Table 2.3. Both tables use rule 6 by having captions that explain (briefly) what they show. Captions should not be used to write essays. If you need to give a long explanation of a table, then put it in the text of your document, not in the caption.

The numbers in each cell are quite disparate, and hence, we already have evidence that the genotype frequencies will be quite different. However, this also makes it quite hard to see what the total sample size is in each group. Row sums are more appropriate here because we are comparing between race codes. Once we have added the row sums it is very easy to see (in Table 2.4) that the sample (database) sizes are approximately the same for each race code. Hence,

	Af. American	Caucasian	SW Hispanic
AA	1	19	9
AB	28	24	20
AC	6	60	48
BB	117	4	15
BC	46	29	62
CC	7	63	54

TABLE 2.2: Count table of race code by genotype for the Gc locus

	AA	AB	AC	BB	BC	CC
Af. American	1	28	6	117	46	7
Caucasian	19	24	60	4	29	63
SW Hispanic	9	20	48	15	62	54

TABLE 2.3: Count table of genotype by race code for the Gc locus

it is easy and valid to make comparisons between the raw counts. However, we stated that our aim was to make comparisons between the proportions. By using proportions, we remove the effects of sample size on our comparisons. It is important, however, to have the sample sizes somewhere as sample size directly affects our perception of how reliable the proportions are. Therefore, we will alter our table so that it shows the proportions, but the final column will contain the sample sizes. Table 2.5 shows the application of the "two busy digits" rounding rule. The rule is essentially to round to the two digits that are conveying most of the information. The motivation for this rule is that by rounding to these two digits we are able to make comparisons more easily than if we had full precision. Of course, these rules are just guidelines and should not be applied blindly. If the aim of presenting tabular data is so that someone else can use your data, then you might reconsider altering the precision so that the loss of information is minimized. However, if the aim is presentation and explanation, then the two busy digit rule serves us well. From Table 2.5, we can now see easily that the African American sample is markedly different from the Caucasian and SW Hispanic population. In particular, many African Americans in the sample carry the BB genotype. The Caucasian and SW Hispanic samples, on the other hand, have very few people with the BB genotype.

2.5.4.1 Comparing two proportions

A natural question that often arises from tables of proportions is "Are the proportions in these two groups (statistically) different?" This question can be easily answered if the two groups are independent. For example, we can perform a statistical test or construct a confidence interval that compares the proportion of African Americans with the BC genotype to the proportion of SW Hispanics who have the BC genotype (or any other genotype) because we can reasonably assume that the African American and Hispanic databases

	AA	AB	AC	BB	BC	CC	Sum
Af. American	1	28	6	117	46	7	205
Caucasian	19	24	60	4	29	63	199
SW Hispanic	9	20	48	15	62	54	208

TABLE 2.4: Count table of genotype by race code for the Gc locus

	AA	AB	AC	BB	BC	CC	n
Af. American	0.00	0.14	0.03	0.57	0.22	0.03	205
Caucasian	0.10	0.12	0.30	0.02	0.15	0.32	199
SW Hispanic	0.04	0.10	0.23	0.07	0.30	0.26	208

TABLE 2.5: Frequency table of genotype by race code for the Gc locus

are independent samples from those populations. However, we cannot, for example, compare the proportion of Caucasians with an AA genotype to the proportion of Caucasians with a BB genotype because these quantities are dependent. To illustrate this, imagine for a moment that there are only three possible genotypes, AA, AB, and BB. If we know the proportions for any two of these three, then we always know the third because the proportions sum to one. There are other ways of comparing proportions in this situation, such as odds, but we cannot look at simple differences.

2.5.5 Comparing groups

When we have one categorical variable and one continuous group, then our aim is usually the comparison of groups. We use the information in the categorical variable to divide the continuous variable into groups. There are three main features that we are interested in comparing amongst the groups. These are

- Location or center
- Scale or spread
- Shape or other features

2.5.5.1 Measures of location or center

Our initial interest is usually the location of the groups. That is, we are interested in using the data we have collected to answer questions like "Is the new method better than the old method?" or "Have these samples come from the same source?" Our measures of center are usually statistics like the mean or the median, or occasionally the mode. The **mode** is the most common value in a sample. If two samples are from the same source, then we might reasonably expect them to have a similar mean, median, or mode. Which statistic is the best measure of center? Unfortunately, the answer is context dependent. The mean is a good descriptor of the center of a sample when the sample is unimodal and symmetric. Unimodal literally means "one mode." If a sample is unimodal, then a histogram or a density estimate of the data will show one major peak. Symmetric means that there is an equal amount of data above and below some central value. Figure 2.1 shows the different descriptions of shape that statisticians use.

If your data is strongly multi-modal, then a single mean or median might be an inappropriate or insufficient summary of the data. For example, Figure 2.2 shows refractive index (RI) measurements made on glass fragments recovered in New Zealand case work. The dotted line is the mean of all 2,656 observations. However, it is clear from Figure 2.2 that there are multiple modes in the data. This would be expected by readers familiar with forensic glass evidence. Very crudely, glass with an RI of around 1.516–1.518 is associated with building glass (glass you would find in the windows of a house or a store), whereas glass with an RI around 1.520–1.523 is vehicle glass. Glass with a very low RI is likely to be a borosilicate like Pyrex®. Glass with a high RI is glass with additional lead such as lead crystal.

The mean is usually not an appropriate choice when the data are highly skewed or contains outliers. We have not discussed outliers, but for the time being we can regard them as a small set of observations that are quite different from the bulk of the data.

Figure 2.3 shows histogram of 150 refractive index measurements made through a piece of float glass including the float surface. The float surface and near float surface has a markedly different and more varied RI than glass taken from the bulk (central) stratum. Although the mean is a legitimate statistic to use, you can see from Figure 2.3 that it does not coincide with where the bulk of the data lie. This is because the mean is badly affected by the upper tail of the histogram. The median, on the other hand, is less affected by these measurements, and therefore does a better job of describing the data. In general, the median is a better measure of center for skewed data than the mean.

2.5.5.2 Measures of scale or spread

There are three common measures of scale or spread. These are the **range**, the sample standard deviation, and the **interquartile range**, or **IQR**. The range of the data is simply the distance between the minimum and maximum values. It is rarely expressed this way, however, and one usually gives both

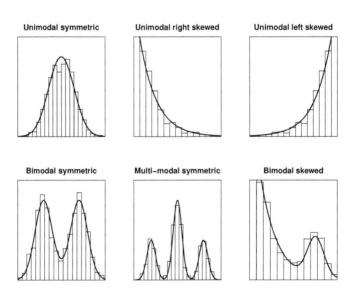

FIGURE 2.1: Descriptions of distributional shape

the minimum and maximum when asked for the range. While the range does describe the complete spread of the data, it is not a good measure of spread

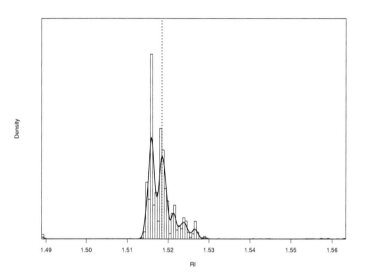

FIGURE 2.2: Refractive index (RI) of glass from New Zealand case work

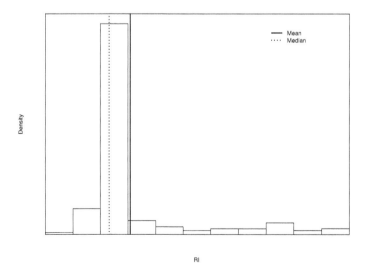

FIGURE 2.3: Refractive index of float glass, including the float surface

in general. In particular it is very sensitive to outliers or errors in the data. Table 2.6 shows the refractive indices of 10 fragments of glass recovered from a suspect in a case I was involved in. I have shown the RIs to four decimal places so that you can see how the numbers vary. The range of this data is 1.5189 to 1.5230. However, inspection of the data shows that the majority of measurements lie around 1.5196 to 1.5199. In this data, there is good reason to believe that the fragment with the highest RI (1.523) is in fact from another source or perhaps from the float surface. As a point of reference, a typical standard deviation for bulk float glass from a single source is in the order of 10^{-5}. An alternative measure is the sample standard deviation. The standard deviation is also sensitive to outliers, but not nearly as much as the range. The standard deviation for the data in Table 2.6 is 0.001. An extremely robust measure of spread, which sometimes can be too crude, is the interquartile range. The interquartile range is simply the distance between the upper and lower quartiles (the 0.25 and 0.75 quantiles). The IQR tells us about the central 50% of the observations. The IQR for the data in Table 2.6 is 0.0002, ($= 2 \times 10^{-4}$). While this is still larger than 10^{-5} it is a much more reasonable measure of the spread in the data. This statistic can be a little crude in that it "throws away" 50% of the data. However, there is no reason we need to stick with the central 50% of the data. We can use any quantiles we like. For example, the central 80% of the data lies between the 0.1 quantile and the 0.9 quantile. The **interquantile distance** is defined as

$$IQR(\alpha) = \widehat{q}_{1-\alpha/2}(x) - \widehat{q}_{\alpha/2} \tag{2.5}$$

where $\widehat{q}_\alpha$ is the α empirical quantile or sample quantile. The sample quantiles are calculated from the order statistics using linear interpolation. This statistic describes the range of the central $100 \times (1-\alpha)\%$ of the data.

2.5.5.3 Distributional shape and other features

Once we have described the centers and spreads of the groups, we can comment on the individual features of the groups. The features of interest are usually **distributional shape** and outliers. The terms we use to describe distributional shape are mostly given in Figure 2.1. Sometimes, however, we do not have histograms of each of the groups. We might have plots of the raw data or perhaps box plots of the data by group. It is more difficult to comment on shape with these plots. We will talk about the features of various plots in Chapter 2. What is important to discuss in this section is the

1.5189	1.5189	1.5196
1.5196	1.5196	1.5197
1.5198	1.5198	1.5199
1.5230		

TABLE 2.6: RI of 10 recovered glass fragments

α	Percent	$IQR(\alpha)$
0.1	90%	0.0027
0.2	80%	0.0013
0.3	70%	0.0007
0.5	50%	0.0002

TABLE 2.7: $IQR(\alpha)$ for the recovered RI data and various values of α

concept of outliers. The definition of an **outlier** is an observation that is substantially different from the bulk of the data. An outlier may be caused by errors in the data, but an outlier is not necessarily an error. This distinction is important. Take the data from Table 2.6 as an example. This data set has RI measurements on 10 fragments recovered from a suspect. The mean RI is 1.5199, and the standard deviation of all the data is 0.0012. Even using these figures, which are likely to be inflated, the maximum value 1.523 is approximately 2.7 standard deviations above the mean. In contrast, the minimum value is 0.9 standard deviations below the mean. Therefore, we might suspect that the fragment RI 1.523 is an outlier. Is it an error? Remember these values come from a case. They have been peer checked and presented to the court. The process of measuring RI is semi-automated by machinery and the results are saved directly to the computer, eliminating the possibility of a data transcription error. Therefore, the likelihood that this observation is an error is low. It is more likely that this fragment comes from the float surface of the glass, or the near float surface, which can have a substantially different RI distribution than the glass that comes from the center (of the thickness) of the glass. Should it be included in the subsequent statistical analysis and interpretation? That is quite a different question. There are circumstances where you may choose to remove outliers from an analysis. You may do this because you believe that the outlying data unduly influences the test, or because you have reason to suspect that a measurement was not made correctly. However, in either case, you should always record removal of observations in your article or lab notes (if the intent is not external publication). In this example, I would probably remove this observation from the analysis and treat it as a "group" of non-matching glass for two reasons. The first of the reasons is given above – significantly different from the bulk glass, and if it is from the scene, then it is likely to be from near the float surface. The second reason is the suspect actually had 20 fragments of glass removed from his clothing. This is an extremely large number, even for someone who is associated with crime, and therefore, it is entirely possible that some of these fragments may have come from another source.

2.5.5.4 Example 2.1—Comparing grouped data

The bottle data set contains the elemental concentration of five different elements (Manganese, Barium, Strontium, Zirconium, and Titanium) in samples of glass taken from six different Heineken beer bottles at four different locations (Base, Body, Shoulder, and Neck). Five repeat measurements are made on each sample at each location. The data set has seven variables, which are listed in Table 2.8.

Variable	Description
$Number$	the bottle
$Part$	location of sample on bottle
Mn	the manganese concentration
Ba	the barium concentration
Sr	the strontium concentration
Zr	the zirconium concentration
Ti	the titanium concentration

TABLE 2.8: The variables in the bottle data set

The bottle data set contains a variable *Part*, which tells us which part of the bottle the fragment that was measured comes from (Base, Body, Shoulder, and Neck). In addition, we have measurements on the concentration of zirconium (Zr) for each observation. Therefore, we find ourselves in the situation where we have one grouping (categorical) variable and one continuous variable. It may be of interest to see whether there is a difference in the zirconium concentrations depending on where the sample comes from. We can see from Table 2.9 that each group has an equal number of observations. This can be important because the number of observations directly affects the stability of statistics like the mean and the standard deviation. The number of observations per group is moderate as well, meaning that we can have some confidence in our estimates. I will elaborate on that statement more in later chapters. We are primarily concerned with whether there are any differences in zirconium concentration between the different body parts. Looking at the

	Base	Body	Shoulder	Neck
N	30	30	30	30
Mean	84	86	82	85
Median	84	81	81	86
Std. Dev.	18	14	12	15
IQR	19	16	15	16
Min.	50	60	56	51
LQ	73	78	76	78
Median	84	81	81	86
UQ	92	94	91	93
Max.	126	119	107	110

TABLE 2.9: Summary statistics for Zr concentration by body part

means, we can see that there is essentially no difference between the body parts. The shoulder has a slightly lower mean than the others, but when we factor in the variability in the form of the standard deviation or IQR, we can see that this difference is insignificant compared to the variability within each part. The data appear to be reasonably symmetric. We can see this from the fact that the means and medians are approximately the same for each part. Furthermore, the distance of the quartiles to the medians is roughly the same for each part.

2.5.6 Two quantitative variables

If we have two continuous variables or two discrete variables, then we can explore the relationship between them. Since this section is concerned with basic statistics, we shall restrict ourselves to correlation analysis or the correlation coefficient. The **correlation coefficient** or Pearson product-moment correlation coefficient is potentially one of the most misused statistics. This is primarily because of a misunderstanding of what it measures.

> *The correlation coefficient measures the strength of the **linear** relationship between a pair of quantitative variables.*

To emphasize the distinction being made here, take a look at the plots in Figure 2.4. Most people would say all four plots exhibit a high degree of correlation between the variables. This is not true. Although there are very strong relationships apparent in all four plots, there are strong linear relationships only in plots A and B. In fact, the respective correlation coefficients are A: 0.99, B: −0.99, C: −0.06, and D: −0.06. Perhaps part of this confusion lies in the fact that regression analysis has a quantity called the **(squared) multiple correlation coefficient** or R^2. This quantity measures the linear dependence between the observed values and values predicted by a model. We can see that there is clearly an appropriate model for the data in plots C and D, and therefore, these data sets would have high R^2 values. Further complicating the issue of interpreting correlation coefficients are the effects of outliers. Statistician Francis Anscombe constructed four sets of data known as the *Anscombe quartet* [7]. Each of these data sets has two variables, x and y. In each data set, x and y have the same mean (9 and 7.5), the same standard deviation (3.32 and 2.03) and the same correlation (0.82). If a simple linear regression model is used to describe the linear relationship between x and y, all four data sets produce the same fitted equation $y = 3 + 0.5x$. These data sets are shown graphically in Figure 2.5.

These four plots are obviously very different. Plot A is adequately described by a linear relationship (although there are only 12 points so that conclusion is a little dubious) and therefore the correlation coefficient is sensible. Plot B exhibits significant non-linearity (a quadratic relationship), and

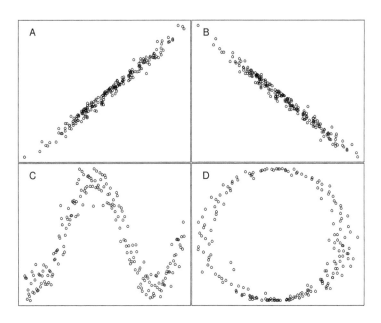

FIGURE 2.4: Four data sets that have highly dependent variables but not necessarily with high correlation coefficients

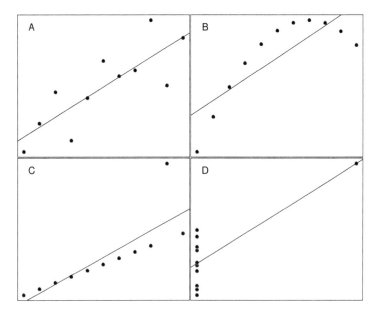

FIGURE 2.5: The Anscombe quartet: all four data sets have the same correlation of 0.82

hence, the correlation coefficient is not providing a sensible summary of the data. Plot C shows the effect of a point that is a long way from the line and is influencing the position of the line. We will see later on that this data point is described as having a large **residual** and high **influence**. The correlation between x and y is being decreased by the point at the top of the plot. Plot D shows a phenomenon known as **anchoring**, and the point to the far right has high **leverage**. The data in this plot in fact exhibit no variation in x except for the point on the right. Without this point the standard deviation of x would be zero, and the covariance between x and y would be undefined since x and y do not "co-vary" (and for mathematical reasons). Situation D may seem extreme; however, it routinely occurs when there are only two values of x that represent two groups in the data. The correlation coefficient is clearly an inadequate description of this situation. The key message of this section is that if you have two continuous variables that you think are related you should plot them first. Why did we not do that? Because many people do not. They simply jump straight in and calculate the correlation coefficient.

It is possible to calculate the correlation between two discrete variables. The behavior of the correlation coefficient in this situation depends on the "resolution" or "granularity" of the variables. By these terms I mean how many different values were observed for each of the variables. For example, one might imagine any situation where you are counting something—the number of fibers found on car seats, the number of pieces of broken glass on the ground, etc. In cases such as these, the possible values are 0, 1, 2, 3,..., up to potentially a very large number. If there are lots of different values in the sample, then there are reasons why we might treat this data as continuous. In these circumstances, the correlation coefficient works in more or less the same way as it does for continuous data because the data are, for all practical intents, continuous. When very few different values are observed, the same sort of care must be taken.

Figure 2.6 shows scatter plots of four different pairs of discrete random variables. The correlation coefficient is performing sensibly for the first three instances. However, the bottom right exhibits the effect of anchoring. The data in this plot are uncorrelated, but the correlation coefficient is inflated by a large influential point. Once again, the take home message is plot the data first if you can.

2.5.6.1 Two quantitative variables—a case study

The bottle data set contains quantitative measurements on the concentrations of five metals: manganese (Mn), barium (Ba), strontium (Sr), titanium (Ti), and zirconium (Zr). If there are dependencies between these elements, then it means that we should not model them separately. Figure 2.7 shows a **pairs plot**, which is essentially a matrix or array of scatter plots. In the upper triangle of the matrix, the pairs plot shows scatter plots of pairs of the concentrations of the elements. The elements being plotted can be determined by following along the row to determine the element on the vertical (y) axis and down the column to determine the element on the horizontal (x) axis. A

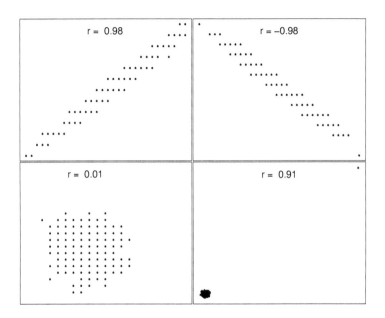

FIGURE 2.6: Correlation between two discrete variables

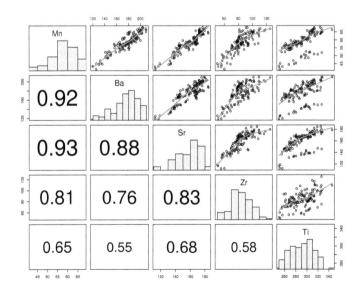

FIGURE 2.7: Pairs plot for bottle data

smoothing line is added to each plot. We will discuss smoothers more later on. The large numbers in the lower triangle of the matrix are the correlations between the pairs of elements. The font size of the numbers is proportional to the magnitude of the correlation. So, for example, we can see that the barium and manganese are highly correlated ($r = 0.92$). The scatter plot shows that the data cluster around a straight line, so the correlation coefficient is an accurate summary of the data. In the case of barium and titanium, however, the correlation coefficient perhaps is not the best summary of the data. We can see from the scatter plot that there a set of points that measures lower in barium. Although barium concentration does appear to increase with titanium concentration, the increase may not be linear. Our conclusion would be that there appears to be a relationship between the barium and titanium concentration, but the exact nature of the relationship requires more exploratory data analysis.

2.5.7 Closing remarks for the chapter

There are two important messages to take from this chapter. Firstly, make sure you understand what the statistics you use are telling you, and secondly, make sure that your statistics are providing your audience with the information they need to understand you. Your use of statistics should not follow the advice of Lewis Carroll—"If you want to inspire confidence, give plenty of statistics. It does not matter that they should be accurate, or even intelligible, as long as there are enough of them"—but rather that of William Watt—"Do not put faith in what statistics say until you have carefully considered what they do not say." Once again, think about what you want to say and if your statistics are saying it. I would recommend reading Denise P. Kalm's presentation *The Minimum Daily Adult* [8], which can be found on the internet. This presentation has good practical advice about the presentation and interpretation of statistics as well as a collection of good quotes, some of which I have used in the preceding paragraphs.

The remainder of this chapter is concerned with installing R and a first tutorial in R.

2.6 Installing R on your computer

The latest version of R can always be downloaded from http://www.r-project.org. On the right-hand side of the web page you will see a link under "Download, Packages" labeled *CRAN*. CRAN is the Comprehensive R Archive Network and is a network of mirrors providing local R downloads for many countries. Choose the mirror that is closest to you. After selecting a mirror site, the CRAN front page will load. At the top of the page under *Download and Install R*, you will see links for Linux, Mac OS X, and Windows. This section will deal with the installation of R under Windows as it is the most common platform in forensic laboratories. There are limited instructions for major Linux distributions and Mac OS X on the web pages for those operation systems. Click on the *Windows* link, then the *base* link on the next page. This will load a page with the top link being to the latest stable version of R. At the time of this writing, this is version 2.10.1 and the download is approximately 32 MB. Click on the link and save the R installer *R-2.10.1-win32.exe*. Once the download has completed, click on the *Run* button if you are using Internet Explorer or double click on the download if you are using Firefox. At this point, simply follow the on screen instructions. It is best to let R install where it wants, and not change any of the default instructions unless you know what you are doing.

Tip 1: Installing R under Windows Vista and Windows 7

Installing R under Windows Vista or Windows 7 can cause you a few small problems if you do not have administrative rights. I strongly urge you to read the R FAQ on this subject `http://tinyurl.com/y9cnp28`

The simplest solution is to tell R to install itself in your home directory. If you have a typical Windows Vista or Windows 7 install, then your home directory will be `C:/Users/loginname` where `loginname` is your login name or account name. For example, on most of my computers, my login and account name is `curran`. Correspondingly, my home directory is `C:/Users/curran`.

Following the instructions in the FAQ will make sure that you do not have issues installing R or installing R packages.

I do customize my own installation in that I like to have the PDF technical manuals and that I prefer the plain text help over HTML or Windows Help (CHM). If you do decide at some point that you would like to change how R was installed, you can simply run the installer again.

Once you have installed R, start it up. You should see something like Figure 2.8, although the dates and version numbers may be different depending on which version of R you have installed.

2.7 Reading data into R

There are many commands for reading data into R. I will describe the two simplest, which are `read.csv` and `scan`.

2.7.1 `read.csv`

The simplest, most portable way to store a data set is in a comma-separated value file (CSV). It is simple because a CSV is a text file, meaning that it needs nothing more than a text editor like Windows Notepad, Emacs, or vi to open it. It is portable, in that it can be easily opened under Linux, Mac OS X, or Windows or even DOS or sent inside an email. A CSV file will usually open directly into Excel or Open Office and the layout of the data will be preserved. The format of a CSV file is very straightforward: each row in a spreadsheet becomes a line in the CSV file; each column within a row is delimited (separated) by a comma and that is it! If your data are stored in Excel or Open Office file formats, there are options to save your data as a CSV file. I will not detail this procedure here as it is operating system and software version specific, but it is trivial. There is also a variety of tools, which let you read your data from other software packages. You can obtain details of these tools by typing

```
> help(package = foreign)
```

at the R prompt.

To illustrate the use of `read.csv`, I have provided a data set that can be downloaded from `http://www.stat.auckland.ac.nz/~curran/bottle.csv`. To get the data, type the URL into the address bar of your web browser and when prompted, save the file to a location that you can find. If you have

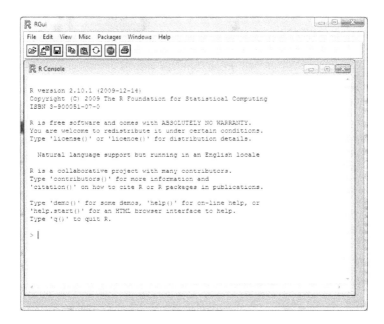

FIGURE 2.8: The R console

Microsoft Excel and you are connected to the internet, then you can start Excel and type the URL into the *Filename:* box of the *File Open* dialog box.

Tip 2: Retrieving tabular data from web pages

A useful, but not well-known, fact about Microsoft Excel is that it can directly read web pages. If you are connected to the internet, then all you need to do is type the URL of the web page (including `http://`) into the *Filename:* box of the *File Open* dialog box. If the web page has data in HTML tables, then these data will be nicely laid out cells on your Excel spreadsheet.

Tip 3: Saving files

Save your files somewhere you can find them, and preferably not on the desktop or in the *My Documents* folder under Windows. The reason for this is that you need to be able to tell R where your file is in the directory structure. This can be quite a complicated if you have saved on the desktop or *My Documents*.

The simplest way to use `read.csv` is to read the data directly off the internet. To do this, you simply type

```
> bottle.df = read.csv("http://www.stat.auckland.ac.nz/
                       ~curran/bottle.csv")
```

This, of course, is not particularly helpful because for the most part you will not be loading data directly from the internet. R has a function called `file.choose` to make it easier for users to locate files. `file.choose` will bring up the standard *Open file* dialog box. To select a file, you simply navigate through the directory structure, choose the file you want and click on *OK*. `file.choose` can be used in conjunction with `read.csv`, e.g.,

```
> bottle.df = read.csv(file.choose())
```

The final way to use `read.csv` is to specify where R can find your file. If the file is in your working directory, which you can set by using the *Change dir...* command from *File* menu or by typing `setwd(choose.dir())`, then you can simply type

```
> bottle.df = read.csv("bottle.csv")
```

If you have saved *bottle.csv* in a directory other than your working directory, then you need to tell R where it is. For example, I saved *bottle.csv* in `c:/users/curran`; therefore, I would type

Basic statistics   31

```
> bottle.df = read.csv("c:/users/curran/bottle.csv")
```

> **Tip 4: File name and file path case sensitivity**
>
> Note that file names and paths are not case sensitive under Windows. If you are using Linux or Mac OS X, you will have to make sure that the case of the file names and paths match.

2.7.1.1 Checking your data has loaded correctly

If read.csv is successful, it does not have any output. That is, it does not say something helpful like *"File loaded"* or *"Read 120 observations."* It says nothing. Therefore, I always recommend that you check your data once you have loaded it with a few simple commands. Firstly, it pays to check that you have the variables you think you should have with the names function. Doing this for the bottle data we get

```
> names(bottle.df)
```

```
[1] "Number" "Part"   "Mn"     "Ba"     "Sr"     "Zr"
[7] "Ti"
```

The second check I recommend is to make sure that you do not have any missing values, or at least no more than you think you should have. This can be done with the sum and is.na functions.

```
> sum(is.na(bottle.df))
```

```
[1] 0
```

R uses the letters NA to represent a missing value by default. If there are no missing values, then this should return zero as it has in our example. One final check is to use R's data editor. This is a very simple spreadsheet-like editor. It can be accessed via the edit command, e.g.,

```
> edit(bottle.df)
```

2.7.2 scan and others

The scan function is a much more low-level function. It is useful either when you have a single variable or when you have an extremely large data set. In its simplest use, for a single variable, scan treats the entire file as containing values for a single variable delimited by spaces.

I have mentioned scan here, not because I expect you to use it regularly, but because it provides an alternate mechanism for reading data. Most data sets that are stored as text files can be read in with `read.csv` or `read.table`. You will occasionally encounter very big data sets, or data sets that have difficult formatting. If your data set is larger than about 10MB, I would recommend you either break it into smaller more relevant parts or consider dealing with it via a database. R supports the Open Database Connectivity (ODBC) protocol through the package `RODBC`. It is relatively simple to query to a data source using this package. If your data set is formatted in an unusual manner, then I suggest you think about performing some cleaning and manipulation beforehand. This can be done with a text editor such as Emacs or with a text processing language like Perl or even R.

2.8 The `dafs` package

All of the data used in this book is in an R package called `dafs`—short for *data analysis for forensic scientists*. To use this package, you need to download it and install it from the internet.

You need to be connected to the internet to install an R package, or you need to have downloaded the package and saved it to disk. The simplest way to install an R package under Windows is to use the **Install package(s)...** option from the **Package** menu. R will ask you to select a *mirror* or location that is geographically close to you, and then ask you to select the package you wish to install. Locate the `dafs` package, click on it to select it, and then click **OK** to install it. Alternatively, you can use the `install.packages` command under Windows and all other operating systems. This requires you to know the URL of at least one CRAN repository. http://cran.r-project.org should work in all circumstances, but if at all possible try and find the URL of a mirror that is closer to you. The code is

```
> install.packages("dafs", repos = "http://cran.r-project.org")
```

This installation procedure only needs to be done once. The package will be updated two to three times a year, so you might repeat the installation procedure if you want to update. Once the package is installed, you need to load it to use it using the `library` function.

```
> library(dafs)
```

> **Tip 5: Using R packages**
>
> R packages must be loaded for every R session. Note that loading a package for use is different from installing it. To use a package we call the `library` function. Each time you quit and restart R you must manually load the packages you want to use.
>
> There is a mechanism for loading a set of packages at the start of an R session which I will not describe here. Information on this can be found by typing `help(.First)` into the R console.

This needs to be done for each R session. That is, if you want to use the `dafs` library, you will need to use the `library` command each time you start R. If you want to update the library, call the `install.packages` function *before* you call the `library` command.

2.9 R tutorial

R has a high initial "frustration factor." The purpose of this tutorial is firstly to provide a set of R commands that you, the reader, should be able to replicate. When you are learning R it is really useful to have a set of examples with output that you can type the commands and get the same results. This should give you some confidence that R is actually working correctly. The other purpose is to get some practice with R. The more you use R, the more familiar you will be with its commands and some of its "quirks."

2.9.1 Three simple things

There are three basic pieces of information that I find can help ease much of the frustration with R. These are

1. R is case sensitive. If R gives you the `Error: object 'x' not found` where `'x'` is the name of the variable or command you think have typed you should check capitalization and spelling.

2. If you are using R in the Windows console (and most of you will be), then the up arrow key (↑) will let you get back to the last line you have typed. This is very useful if you make a spelling mistake or miss a bracket or a quote.

3. You can get help on any R function by using the `help` command, or its shorthand version `?`. For example, you can type `help(mean)` or `?mean`

to get help on the mean function. You can also use the `help.search` command to find help files or the names of commands you have forgotten. For example, if I could not remember that the standard deviation of a variable can be calculated in R with the `sd` command, I could type `help.search("standard deviation")`, which would return the following output.

```
Help files with alias or concept or title matching
'standard deviation' using fuzzy matching:


fda::sd.fd              Standard Deviation of Functional Data
nlme::pooledSD          Extract Pooled Standard Deviation
stats::sd               Standard Deviation


Type '?PKG::FOO' to inspect entry 'PKG::FOO TITLE'.
```

I can now see from the output that the command I want is `sd`. I can see this in the line `stats::sd`. The `stats::` tells me that `sd` is in the `stats` package, which R loads by default.

2.9.1.1 Tutorial

1. Create a vector x which has the values 11, 9.3, 11.5, 8.9, and 11.2

   ```
   > x = c(11, 9.3, 11.5, 8.9, 11.2)
   ```

   Note that R does not (usually) pay any attention to white space, so it does not matter whether you put spaces before or after the commas or brackets. The = sign is called the assignment operator. When you look at the R documentation you will see that many examples use <- instead of =. This is because the traditional assignment operator in R is <-. There are very few circumstances, however, where it matters whether you use the = sign or <-. I will use = wherever possible in this book. The c function is used to create vectors. The origins of the name of function are hidden in the mists of time. I like to think of it as the concatenation function because it joins elements together.

2. Check the value of x by printing it out. To print out the value of any R variable (or object) simply type its name and hit return

   ```
   > x
   
   [1] 11.0  9.3 11.5  8.9 11.2
   ```

3. Summary statistics in R.

a Calculate the mean, median, and standard deviation of x using the mean, median, and sd functions.

```
> mean(x)
[1] 10.38
> median(x)
[1] 11
> sd(x)
[1] 1.190378
```

b Calculate the upper and lower quartiles, and the interquartile range of x using the quantile, IQR, and diff functions.

```
> quantile(x, 0.25)
25%
9.3
> quantile(x, 0.75)
 75%
11.2
> IQR(x)
[1] 1.9
> diff(quantile(x, c(0.25, 0.75)))
75%
1.9
```

The last two commands are just two different ways of calculating the interquartile range.

c Calculate the interquantile distance with $\alpha = 0.2$.

```
> alpha = 0.2
> quantile(x, 1 - alpha/2) - quantile(x, alpha/2)
 90%
2.32
> diff(quantile(x, c(alpha/2, 1 - alpha/2)))
 90%
2.32
```

The last two commands show you two different ways of calculating the interquantile distance. Of course, you can also just substitute the values of $\alpha/2$ and $1 - \alpha/2$ directly into the commands as well, but it is more readable this way. Ignore the 90% label on the answer—that is just a side effect of using the quantile command.

4. Sort x into ascending order, using the `sort` function and store it. This means that x will permanently sorted into order.

```
> x = sort(x)
> x
```

```
[1]  8.9  9.3 11.0 11.2 11.5
```

If you had just typed `sort(x)`, then this would have printed out x in ascending order but not changed it. By typing `x = sort(x)`, we have *reassigned* the sorted values of x to x.

5. Extract the 1st, last (5th), the 3rd, 2nd, and 4th elements of x in that order, and elements 1 to 3.

```
> x[1]
```

```
[1] 8.9
```

```
> x[5]
```

```
[1] 11.5
```

```
> x[c(3, 2, 4)]
```

```
[1] 11.0  9.3 11.2
```

```
> x[1:3]
```

```
[1]  8.9  9.3 11.0
```

6. Read in the bottle data and store it in a data frame called `bottle.df`. Find out the names of the variables in `bottle.df`. You will need to be connected to the internet to complete this task.

```
> bottle.df = read.csv(
    "http://www.stat.auckland.ac.nz/~curran/bottle.csv")
> names(bottle.df)
```

7. Calculate the mean concentration of manganese in all the bottle samples in `bottle.df`. There are several ways to do this. Try them all.

```
> mean(bottle.df$Mn)
```

```
[1] 58.19583
```

```
> mean(bottle.df["Mn"])
```

```
      Mn
58.19583
```

```
> with(bottle.df, mean(Mn))
[1] 58.19583
```

8. Calculate the mean concentration of barium for each bottle in `bottle.df`.

```
> sapply(split(bottle.df$Ba, bottle.df$Number),
    mean)
      1       2       3       4       5       6
181.920 183.310 172.555 187.150 138.730 189.880
> with(bottle.df, sapply(split(Ba, Number), mean))
      1       2       3       4       5       6
181.920 183.310 172.555 187.150 138.730 189.880
```

I realize that the commands in this task look unduly complex. However, if we break them down, we can see how it works. There are two key functions, `split` and `sapply`. `split` breaks up one vector on the basis of the groups that exist in another. Therefore, in our example, we know that `bottle.df$Number` tells us which bottle we are dealing with, and `bottle$Ba` tells about the barium concentration of each bottle. The split function returns an R `list` where each component of the list is the barium concentrations of one bottle. Try this out.

```
> split.data = split(bottle.df$Ba, bottle.df$Number)
> names(split.data)
[1] "1" "2" "3" "4" "5" "6"

> split.data["1"]

$`1`
 [1] 170.7 166.2 184.2 170.5 185.2 180.5 170.7 189.6 180.0
[10] 191.6 186.7 170.3 192.5 181.5 195.3 182.9 173.8 199.9
[19] 173.5 192.8
```

The function `sapply` "applies" an R function to every element of a list. Hence, in this example, the mean of each group in the list is taken.

```
> sapply(split.data, mean)
      1       2       3       4       5       6
181.920 183.310 172.555 187.150 138.730 189.880
```

The `with` command means that we can use the variable names in a data frame directly without having to type the data frame name and the `$` sign first. For example, we saw

```
> with(bottle.df, mean(Ba))
```

is equivalent to

```
> mean(bottle.df$Ba)
```

It is debatable in this example, which is more efficient, but using `with` when you need to use more than one variable from the data frame is generally very sensible.

9. (**Advanced**) This last task is really just to have a bit of fun and show how extensible R is. Hopefully, you can see that if you replace `mean` with `sd` in the previous task you will get the standard deviation of each group. It would nice to be able to get the number of observations, the mean, the median, the standard deviation, and perhaps the sum of each group all in one go. To do this we can write a little R `function` that does this for a single vector, and then use it with `sapply`.

```
> mySummary = function(x) {
    resultVector = c(length(x), mean(x), median(x),
        sd(x), sum(x))
    names(resultVector) = c("N", "Mean", "Median",
        "Std Dev.", "Sum")
    return(resultVector)
  }
> sapply(split.data, mySummary)
                   1           2           3           4
N          20.000000    20.00000    20.00000    20.00000
Mean      181.920000   183.31000   172.55500   187.15000
Median    182.200000   183.75000   173.25000   186.35000
Std Dev.    9.819722    11.10419    10.38032    13.32795
Sum      3638.400000  3666.20000  3451.10000  3743.00000
                   5           6
N           20.00000    20.00000
Mean       138.73000   189.88000
Median     141.65000   194.15000
Std Dev.    10.20589    15.36916
Sum       2774.60000  3797.60000
```

2.9.2 R data types and manipulating R objects

We have not explicitly talked about R objects or variables yet. That is because this is not a book on R programming. However, there are some basics that you will need to know in order to understand some of the things that are done in this book. A **data type** is a way of describing what kind of data a computer programming language can store. There are three fundamental data types that we will encounter on a regular basis. I will briefly describe each of these data types in turn:

numeric —the numeric data type is used to represent numbers. R does not (normally) distinguish between integers and real numbers. It (normally) treats all numbers as being real. That is, there is no difference between 1.0 and 1 in R.

string literal —a string literal, string, or character string, is a set of one or more ASCII characters enclosed between quotes. E.g., "x", "James" and "123" are all string literals. Note that because the last example "123" is in quotes, it is treated as a string literal rather than a numeric type by R.

logical —the logical data type is used to represent *true* and *false*. R reserves two special words TRUE and FALSE to represent these states. There are also shorthand versions T and F.

Tip 6: Quotation marks in R

In most circumstances R does not differentiate between single quotation marks ' and double quotation marks ". I have tried, for the most part to use " where possible throughout the book, but the only place where I am aware it will make a difference is if you wish to print quotation marks. The easiest way to deal with this situation is to **escape** the quotation marks by typing a backslash (\) before each quotation mark.

A **data structure** is used for storing one or more data types. An object is a data structure possibly with extended functionality. We will not spend any more time discussing the differences between objects and data structures, but rather examining the four main data structures that you will use in R. These are vector, matrix, data.frame, and list. I will briefly describe each of turn.

vector —an R vector is a set of one or more numerics/numbers, character strings, or logicals. The elements of a vector must all be the same. That is, a vector must contain all numbers, or all characters, or all logicals. If you use a mix of data types in the specification of a vector, then R will convert all the elements to one type.

matrix —an R matrix is a two-dimensional rectangular data structure consisting of rows and columns of data. The elements of a matrix must all be of the same type like a vector. That is, it is not possible to have a column of names followed by a column of ages for example.

data.frame —an R data.frame is R's representation of a statistical data set. The columns of a data frame represent the variables in the data set. The rows of a data frame represent the observations. The columns of a data frame do not need to be of the same data type. However, the elements within each column do need to be of the same data type and all the columns need to be of the same length.

list —an R list can be used to hold one or more elements which may be single values, or other R data structures or objects. E.g., l = list(a = "A", x = c(1,2,3), y = list(T,F,T)) produces a list l with elements a—a string literal "A", x—a numeric vector of length 3, and y—a list of logicals of length 3. The elements of a list do not need to be of the same length or even of the same dimension. A list is the data structure equivalent of a shopping cart—it will hold just about anything.

2.9.2.1 Tutorial

1. Vectors can be created with the c function.

```
> x = c(1, 3, 5, 7, 9, 11)
```

There are some functions that make it easy for you to create patterned vectors. The three most common tasks are: making a vector that has a sequence of integers from a to b; making a vector that has evenly spaced real numbers from a to b; and making a vector that has a certain pattern repeated over and over in it. There are many ways to achieve these goals, but we will use the : operator, the seq command, and the rep command. The : operator has syntax a:b and will produce a sequence of numbers from a to b. a and b do not need to be integers, but it is uncommon to use : for non-integer values of a and b. Some examples are

```
> x = 1:10
> x

[1]  1  2  3  4  5  6  7  8  9 10

> y = -3:3
> y

[1] -3 -2 -1  0  1  2  3

> z = -1.5:5.6
> z

[1] -1.5 -0.5  0.5  1.5  2.5  3.5  4.5  5.5
```

The `seq` command usually takes two arguments with an optional third. The arguments are `from`, `to` and either `length` or `by`. Some are examples of use are

```
> x = seq(-1.5, 0.6, length = 10)
> x
```

```
[1] -1.5000000 -1.2666667 -1.0333333 -0.8000000 -0.5666667
[6] -0.3333333 -0.1000000  0.1333333  0.3666667  0.6000000
```

```
> y = seq(-1.5, 0.6, by = 0.1)
> y
```

```
 [1] -1.5 -1.4 -1.3 -1.2 -1.1 -1.0 -0.9 -0.8 -0.7 -0.6 -0.5
[12] -0.4 -0.3 -0.2 -0.1  0.0  0.1  0.2  0.3  0.4  0.5  0.6
```

The `rep` command takes two arguments: the pattern you wish to repeat and the number of times you wish to repeat it. Some examples of use are

```
> x = rep(1:3, 2)
> x
```

```
[1] 1 2 3 1 2 3
```

```
> y = rep(1:3, rep(2, 3))
> y
```

```
[1] 1 1 2 2 3 3
```

```
> z = rep(LETTERS[1:5], 1:5)
> z
```

```
 [1] "A" "B" "B" "C" "C" "C" "D" "D" "D" "D" "E" "E" "E" "E"
[15] "E"
```

2. The individual elements of a vector can be accessed using the square brackets [ ]. The index, which goes in between the square brackets, can be a single number, a logical value, a vector of positive or negative integers (not necessarily in order) or a vector of logical values. For example,

```
> x = sample(1:10, 10)
> x
```

```
[1]  3  6  5  4 10  8  7  9  2  1
```

```
> x[1]
```

```
[1] 3

> x[5]

[1] 10

> x[c(7, 8, 1)]

[1] 7 9 3

> x[-c(7, 8, 1)]

[1]  6  5  4 10  8  2  1

> x < 5

 [1]  TRUE FALSE FALSE  TRUE FALSE FALSE FALSE FALSE  TRUE
[10]  TRUE

> x[x < 5]

[1] 3 4 2 1
```

3. There are many ways to make a matrix. One way is to use the `matrix` command. However, I will not focus on the construction of matrices here, but rather how to access the elements of matrices. The same syntax can be used with a `data.frame`. When we use the `read.csv` function the data set is stored in a data frame.

   To extract or refer a particular element from an R `matrix` or a `data.frame`, we need to provide R with a row index, and a column index. To obtain the ith element from the jth column, we use the syntax $[i,j]$ after the name of the matrix or data frame. For example, `bennett.df` is a data frame with 10 rows and 49 columns corresponding to 10 RI measurements from 49 different locations in a windowpane. `bennett.df[2,10]` will return the 2nd element of the 10th column from `bennett.df`. If we omit one of the indices, then the whole row or whole column is returned depending on which index we miss. For example, `bennett.df[,10]` will return all 10 elements of column 10 and `bennett.df[3,]` will return all 49 elements from row 3. We can obtain a set of rows or columns by providing R with a set of row indices or column indices. Try these examples (I have not included the output here because it takes too much space. Instead I have commented the R code with an indication of what you should expect to see if you carry out the example correctly.):

   ```
   > data(bennett.df)
   > ##  returns columns 1 to 10 from bennett.df
   > bennett.df[,1:10]
   ```

```
> ##   returns rows 1, 5 and 7
> bennett.df[c(1,5,7),]
> ##   returns the first 5 rows from the first ten columns
> bennett.df[1:5,1:10]
```

4. There is an additional way of extracting a particular variable (or column) from a data.frame using the $ operator. You have seen many examples of this already with the bottle data set.

```
> data(bottle.df)
> bottle.df$Zr
> bottle.df$Part
```

5. A list can be created using the list command. However, it is more important to know how to access the elements of a list. There are three ways of doing this. Firstly, the elements of a list can be access in numerical order using the double square brackets, [[ ]]. The index is a non-zero positive integer, e.g.,

```
> l = list(c(1, 2, 3), LETTERS[4:7])
> l

[[1]]
[1] 1 2 3

[[2]]
[1] "D" "E" "F" "G"

> l[[1]]

[1] 1 2 3

> l[[2]]

[1] "D" "E" "F" "G"
```

Secondly, if the elements of the list have names, then the $ operator can be used in the same way as you would for a data frame. For example,

```
> l = list(x = 1:3, y = 4:6)
> l$x

[1] 1 2 3

> l$y

[1] 4 5 6
```

Finally, if the list elements have names, the double square brackets or the $ operator can be used in conjunction with the names in quotes. This syntax is useful when the list element names have spaces in them. E.g. taking l from the previous example, we could have typed

```
> l[["x"]]
```

[1] 1 2 3

```
> l$"y"
```

[1] 4 5 6

Chapter 3

Graphics

I'm no mathematician, so I'm stuck with the graphic representations – Hugh Hopper, Musician.

3.1 Who should read this chapter?

Simply put, everyone should read this chapter. A good graph is almost better than any statistical analysis. Drawing a good graph is a skill. This chapter covers good and bad habits in drawing a graph and explains how R can be used to make high-quality publication ready graphs.

3.2 Introduction

In this chapter we will discuss a little bit of the philosophy behind creating graphical representations of data and the practical side of producing high quality statistical graphics. By "high quality" I mean graphics that are suitable for publication.

3.2.1 A little bit of language

In this chapter I will use the words *drawing*, *graph*, *(statistical) graphic*, and *plot* interchangeably. I will not use the word *chart*, which has strong associations with Microsoft® Excel®.

3.3 Why are we doing this?

A fundamental question I think many scientists fail to ask themselves is "Why am I drawing this graph?" This question underlies our aims for presenting data in a graphical form. The answer should be "Because I want the reader or the audience to be able to easily understand the point that I am making with this data." There are, of course, a lot of subsequent questions like "What is the most important feature?" or "What do I think is the most interesting thing about this data?" but they all basically come back the principal question. We approach this section with that purpose in mind—to be able to produce the graph that best displays to the intended audience the information or point we want to make. As I stated earlier, this question often gets overlooked and scientists often include a graphic because they think they should or, even worse, because the software they are using allows them to do so easily. A little bit of forethought about the points that you wish to highlight and the intended audience can improve the quality of your graphics greatly.

3.4 Flexible versus "canned"

Software systems that produce graphics generally come in two flavors that I like to call "flexible" and "canned." Canned graphics programs are those where the user is restricted to a set of predefined graphs, such as a scatter plot, a bar graph, etc., with little or no scope to alter them. This is problematic because it constrains the user to producing the graphic the program is capable of rather than perhaps the graphic that the user wants or needs. A fully flexible graphics program allows the user to change absolutely everything. Such flexibility usually comes at the cost of complexity. The more flexible a program is the more complex it becomes. A useful compromise is to offer the "canned" graphs as a simple default, but to provide the flexibility behind the scenes. This is the way that R and many other statistical software packages work.

3.5 Drawing simple graphs

The form or type of the data can guide us in our choice of the most appropriate plot. We should, of course, treat this information simply as a guide and defer to our principle of choosing the graph that best meets our needs if necessary.

We will primarily discuss grouped data or data sets with two or more variables. However, there are a number of circumstances where it is appropriate to discuss the display of information from one variable, and we shall deal with those accordingly.

3.5.1 Basic plotting tools

3.5.1.1 The bar plot

The **bar plot** is used to represent frequency table or count table data. As the name suggests, a bar plot consists of a series of bars, which may be displayed horizontally or vertically. There is (usually) one bar for each category, and the heights or lengths of the bars are proportional to the sample frequencies of counts falling into each category. The widths of the bar plot have no meaning, and there are usually gaps between the bars.

3.5.2 The histogram

The **histogram** is used to show the distributional shape of a continuous variable. A histogram is constructed by first dividing the range of the data in a set of class intervals or breaks. A frequency table is then constructed by counting the number of observations that fall into each of the class intervals. A histogram consists of a series of bars, one for each class interval, where the *area* of the bar is proportional to the number of observations falling into each class interval. The heights *and* the widths of the bars are important because the area contained in the bars, and not height, is used to represent the frequency of the observations falling into each class interval. Usually the widths of the bars are the same, in which case the heights of the bars are proportional to the frequencies. Some software packages use the bar plot routine to produce both bar plots and histograms, which occasionally leads to poor plots. Histograms are sensitive to the choice of the class interval widths.

3.5.3 Kernel density estimates

A **kernel density estimate** (**KDE**) is used to show the distributional shape of a continuous variable. It can be regarded as a smoothed histogram. A KDE is constructed by placing a smooth kernel with a fixed bandwidth at a series of points that span the range of the data, and then taking the average of the density heights for the data as the height of the density at this point. I realize that sounds quite technical and may be better illustrated with an example. Bennett et al. [9] was interested in the spatial variation of refractive index across a pane of glass. As part of that research, 490 RI measurements were made from the same source of glass. Figure 3.1 shows the construction of a KDE on a sample of 50 of these RI measurements. I have chosen 50 so that it is easy to see what is happening. The range of the data has been divided into nine equally spaced intervals. The choice of nine intervals is arbitrary, as is the equal spacing. Starting at the top left, we center a normal curve or density on the start of the first interval. The standard deviation of this normal density is the **bandwidth**. The bandwidth of a kernel density estimate controls the smoothness. In general, the larger the bandwidth the smoother the density estimate. We then take the average height of the density for all 50 data points. The average is depicted as the large black dot. The point is low because there is not much data in this region. We repeat this process for the remaining eight intervals (going from top left to bottom right). When we have finished, the resulting sequence of joined black dots is the kernel density estimate for the data. As a point of comparison Figure 3.2 shows the kernel density estimates, and histogram, of all 490 RI values. The three plots show the effect of changing the bandwidth. The plot on the far left has half the default bandwidth, and the plot on the right has double the default bandwidth. The central plot uses the default bandwidth calculated by R. As the bandwidth increases the density estimate becomes smoother and smoother, and generally less informative. This figure makes it easy to see where the "smoothed histogram" comes from.

Three common choices for kernel functions are shown in Figure 3.3, although the normal or Gaussian kernel is the most commonly used. The ker-

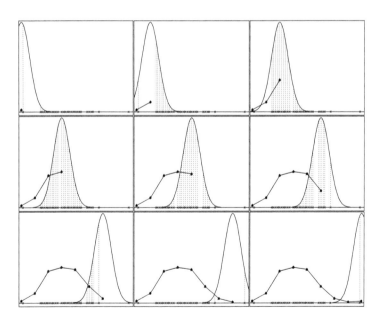

FIGURE 3.1: Constructing a kernel density estimate for 50 RI measurements from the same source

nel density estimate can have an advantage over a histogram in that it allows

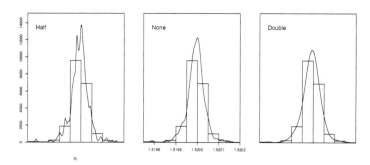

FIGURE 3.2: Histogram and KDE of all 490 RI values

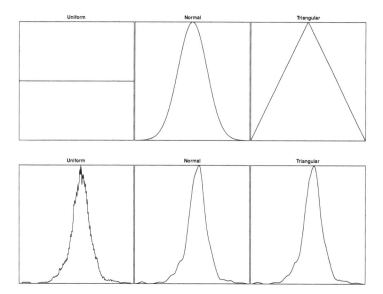

FIGURE 3.3: Three typical kernels and their effect on the density estimates for the RI data

for the possibility we may not have observed the entire range of the data. This is important when we are assessing the value of evidence. Kernel density estimates can be quite sensitive to the choice of bandwidth.

3.5.4 Box plots

Box plots or **box and whisker plots** can be a good tool for quickly comparing grouped data with respect to a continuous variable. Box plots provide a visual display of a set of summary statistics.

Figure 3.4 shows the components of a box and whisker plot. The box is drawn from the lower quartile (LQ) to the upper quartile (UQ). A line is drawn across the box to indicate the median. The **whiskers** are drawn approximately 1.5 times the interquartile range (IQR) from the lower and upper quartiles and then "shrunk back" to the nearest data points. The **outside points** are simply points that are outside the range of the whiskers. They can be outliers, but they are not necessarily outliers. The position of the median line inside the box and the relative lengths of the whiskers tell us about symmetry or skewness of the data. That is, if the whiskers are roughly of equal length and the median line is in approximately the middle of the box, then the data is symmetric. If one of the whiskers is very long and one very short, then the data is skewed. Box plots are good for getting a quick overview of the positions and spreads of grouped data. However, box plots are not sensible when there are relatively few observations per group (< 20).

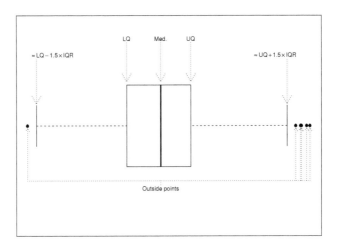

FIGURE 3.4: The components of a box and whisker plot

3.5.5 Scatter plots

Scatter plots, perhaps the most familiar type, are useful when we have two quantitative variables. The scale of the horizontal axis represents the range of one variable and the scale of the vertical axis represents the range of the others. The axes may be on a **linear scale** (e.g., 0, 1, 2, 3, ...) or a **logarithmic scale** (e.g., 1, 10, 100, ...) depending on the nature of the data being displayed.

3.5.6 Plotting categorical data

The primary tool for plotting categorical data is the **bar plot**. There is a reasonable argument that bar plots of single categorical variables are a waste of space and ink. For example, we saw in data on the frequencies of genotypes at the Gc locus for various race codes in Table 2.4 (p. 16). Figure 3.5 shows a bar plot of the genotypes proportions regardless of race code and Table 3.1 shows the same information numerically. There are two questions you should ask yourself. Firstly, "Is the plot easier to understand than the table?" and secondly, "Does the plot convey any information that the table does not?" I believe the answer to both of these questions is no. If you do think that a bar plot is the best way to make your point or display some data that you have collected, then there are two additions that can help increase the utility of the plot. These are the cumulative frequency curve (or ogive) and the bar

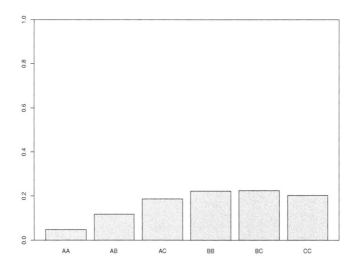

FIGURE 3.5: Bar plot of the genotype proportions for the Gc locus regardless of race code

proportions. Both of these features make it easier for comparisons among the relative frequencies to be made. It can also make sense to order the categories in descending order. A plot showing the categories in decreasing order and with a cumulative frequency line is sometimes called a **Pareto chart**. A Pareto chart is used to identify the most important (or largest) categories. Figure 3.6 shows the use of these features. The data are a color categorization of 12,149 fibers recovered from human heads [10]. The cumulative frequency makes it easy to see that more than 95% of the fibers fall into the first three categories. The bar proportions make it easier to read the contributions of the very small categories to the total.

Genotype	Frequency
AA	0.047
AB	0.118
AC	0.186
BB	0.222
BC	0.224
CC	0.203

TABLE 3.1: Genotype proportions for the Gc locus regardless of race code

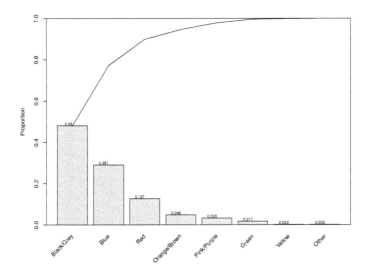

FIGURE 3.6: Different colors of fibers found in human hair ($n = 12,149$)

3.5.6.1 Plotting groups

One situation where bar plots can be useful is comparing proportions between groups. There is a limit to this, however, in that the plots are only sensible when the number of categories within each group or number of groups is relatively low. That is, if you have more than five to six groups or more than four or five categories to compare, then a bar plot may not be sensible. Figure 3.7 shows a bar plot of the genotype proportions for each race code in the FBI sample. I have used different (grayscale) colors to differentiate the groups, and I have added a legend to aid interpretation.

There are situations where we will want to perform statistical tests of significance for the difference between the proportions falling into each category. It is possible to add "error bars" or confidence intervals around the proportions on each plot. However, this practice is quite distracting. There is simply too much information in the plot and it is best displayed in other ways. We will explore this in later chapters.

3.5.6.2 Pie graphs, perspective, and other distractions

My aim throughout this book is to emphasize good practice and not spend an excessive amount time criticizing the things I regard as bad practice. For this reason I have shown you how I think simple categorical data should be displayed with bar plots and not with pie graphs. In this section I will cover some of what I consider bad practice and the reasons.

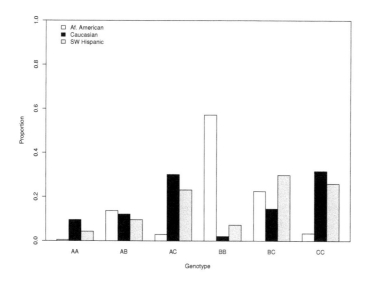

FIGURE 3.7: Genotypes for the Gc locus by race code

Pie graphs are commonly used in the scientific literature and the popular press when the data being displayed are proportions or percentages (of a whole). People use pie graphs because they are "intuitively understood" by most audiences. Most of us understand that the segments of a pie graph represent a proportion of the whole. This is probably because that many of us were taught fractions with pie segments. What is not well recognized is that people are inherently poor at judging relative differences with angles. That is, we are poor at estimating the difference between two proportions when those proportions are represented as angles. Cleveland and McGill [11] carried out a series of experiments in the mid-1980s to assess this explicitly and found that the subjects were twice as likely to make errors in assessing proportions if they used angles (pie charts) versus positions (the tops of the bars). Furthermore, Cleveland and McGill found that accuracy was worse when the proportions were around 0.5 (50%) compared to when the proportions were either very small or very large. As an example of these findings, I have plotted the Gc genotype frequency information as a pie graph and as a bar plot in Figure 3.8. I can see that AA, followed by AB, has the smallest proportion. Past that, however, I cannot distinguish between the remaining categories. By comparison, I can see from the bar plot that AC and CC are similar and BB and BC are the largest. With little extra effort, I can recover the original ranking: AA, AB, AC, CC, BB, BC. You might argue that I am setting up a "straw man" in this example and that normally we would add the proportions to each of the segments in the pie graph. However, if we are going to do

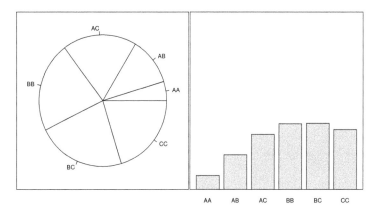

FIGURE 3.8: Pie graph versus bar plot of Gc genotype proportions

that, then I would return to my original question: "What extra information is this plot giving that I cannot get from the frequency table?" This task of comparing angles becomes harder still when we are comparing groups. Figure 3.9 shows the genotypes by race code for the Gc locus. I have deliberately used a separate pie graph for each genotype for two reasons. Firstly, we are interested in comparing the difference between allele frequencies between the race codes. Secondly, this graph is comparable to the bar plot in Figure 3.7 in that the data are grouped by race code.

Adding perspective ("3D") effects to pie charts is a very simple thing to do in some software packages. However, the addition of perspective further reduces our ability to judge angles. That is, because in order to mimic the effect of rotating the plots "in and out of the plane of the page," we have to make the segments in the background of the plot smaller and the segments in the foreground of the plot bigger. The amount of scaling that must be done depends on the angle of rotation. However, any such scaling will, by definition, change the information in the plot. Exploded or divided pie graphs also lead to misleading interpretations. An exploded pie graph is one where the effect of moving the segments apart is added, as though someone has "wiggled the knife around as they cut the pie into segments." The act of moving the segments away from the center of the pie makes the task of comparing angles even more difficult. When this effect is used in combination with a perspective effect, the resulting graphic becomes almost impossible to interpret sensibly.

It is also possible to add perspective or shadow effects into bar plots.

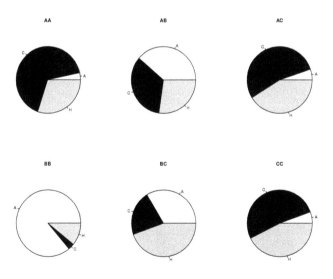

FIGURE 3.9: Using pie graphs for comparison of groups is difficult

Once again, the scaling needed to achieve the effect of perspective introduces distortion into the information the plot is trying to convey. There is sometimes an attempt to counter this distortion by adding horizontal grid lines. However, if you have to introduce counter-measures to remove a "feature" of the graph, then you must ask yourself why you introduced that feature in the first place.

Stacked bar plots are a commonly used method of displaying grouped information on a single bar plot. Figure 3.10 shows the Gc genotype proportions as a stacked bar plot. Cleveland and McGill [11] showed that people were between 1.4 and 2.25 times less accurate with measurements expressed as relative lengths rather than relative positions. This is the case with stacked bar plots. If you wish to compare the frequency of the AB genotype in Hispanics to the frequency in African Americans, then you must compare the length of the gray box with the length of the white box in Figure 3.10. To make this comparison in Figure 3.7, you compare the tops of the boxes.

3.5.7 One categorical and one continuous variable

In this situation we are concerned with the graphical display of the differences (or similarities) in groups of observations with respect to a continuous variable. The way we choose to display this information can depend on the number of observations per group and potentially the number of groups. The tools we will use are the scatter plot, the histogram, the kernel density estimate, and the box plot.

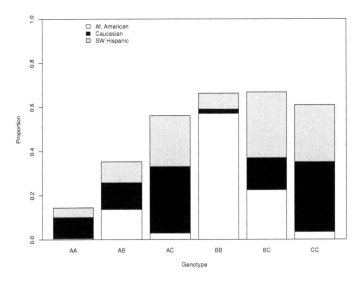

FIGURE 3.10: A stacked bar plot genotypes for the Gc locus by race code

A very simple way to get an initial understanding of grouped continuous data is simply to assign the groups a number from 1 to k, where k is the number of groups. You can then plot the group number on the x-axis and the continuous variable on the y-axis. For example, Newton et al. [12] were interested in the differences in refractive index between the different strata in a piece of float glass. Float glass is made by floating the molten glass on a bed of tin. This achieves the extremely smooth surface seen in post-1950s glass. Newton et al. measured 30 fragments on each of the five strata - float surface (FS), near float surface (NFS), bulk (B), near anti-float (NAFS), and anti-float (AFS). To get an initial feel for this data, we can assign the each stratum a value from 1 (=FS) to 5 (=AFS) and plot RI against this number.[1] This plot is shown in Figure 3.11. It is immediately obvious from this plot that the float surface has a much higher refractive index than the bulk glass and tremendous variability. The anti-float surface also has relatively high variability and appears to have a lower refractive index on average. This plot is useful for getting an initial feel for the data; however, it tells us very little about distributional shape.

3.5.7.1 Comparing distributional shape

Comparing distributional shape between grouped data can be moderately difficult even for small numbers of groups (> 4). Our tools for assessing

---

[1]A convention we will use through this book is that we plot y against x.

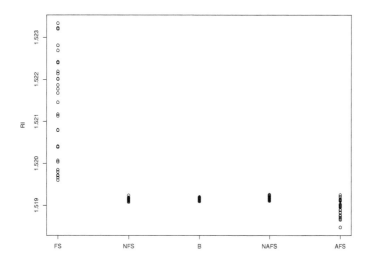

FIGURE 3.11: Refractive index of a sample of float fragments by stratum

distributional shape are the histogram, the kernel density estimate, and to some extent box plots. It makes almost no sense to produce a histogram or a kernel density estimate for fewer than 20 observations, because it really makes no sense to compare distributional shape for fewer than 20 observations. The reason, of course, is that we really do not have enough information in general from such a small sample of data to say anything about the distribution it may have come from.

> **Tip 7: Always compare location and spread on the same scale**
>
> When you are comparing the location and spread of groups with different plots, make sure that they all have the same scale. That is, make sure the axes span the same range and that the plots are the same size. If you do not do this, then comparison can be very difficult.

Figure 3.12 shows histograms with kernel density estimates superimposed for each stratum. The scales are different in each of these plots, as we have already established that there are differences in some of the centers and spreads. There are not many remarkable features in these plots except for perhaps some bimodality in the float surface (FS) group. Box plots are often a good way

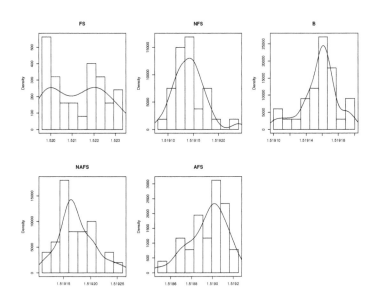

FIGURE 3.12: Kernel density estimates and histograms of the float glass RI by stratum

to rapidly compare grouped data. Figure 3.13 shows box plots of the concentration of zirconium in six different beer bottles. We can very easily see the differences in location and spread between the bottles. Some authors criticize box plots because they "obscure the detail" of the data. That is, one cannot see fine-scale features of the data in a box plot. If this is a concern, then it is trivial in R to plot the data points over the top of the box plots. We will see some examples of this later on where we have a small number of observations per group.

3.5.8 Two quantitative variables

There are three possibilities when we have two quantitative variables: two discrete, one discrete and one continuous, and two continuous. If you have two discrete variables that do not show much variation, then it may be the case that no plot is going to shed light on the data. In situations such as this, you may be better off treating the data as categorical and cross-tabulating it. For example, Found and Rogers [13] looked at the performance of 15 document examiners on a set of 16 questioned signatures from genuine sources and a set of 64 questioned signatures from simulated sources. Examiners were asked to grade the signatures as correct, misleading, or inconclusive. If we are interested in where there is a relationship between the number of misleading signatures in each experiment (genuine versus simulated), then we might plot the number of misleading signatures for each subject. However, this plot would

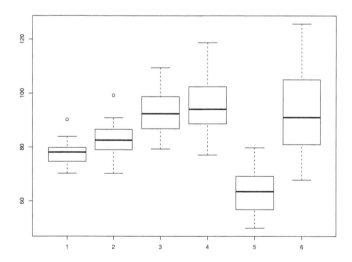

FIGURE 3.13: Zirconium concentration (Zr) in six different beer bottles

not be very informative as even though the data may range from 0 to 16 and 0 to 64 for each examiner on each question the observed values range from only from zero to two in each set of experiments. That is, not many signatures were classed as misleading in either experiment. It may be better to produce a two-way table of this data because of the lack of variability. Table 3.2 shows this table. We can see that there is not much correspondence between the two experiments. Furthermore, there were fewer classifications of "misleading" in the simulated group than in the genuine group.

In every other situation, a scatter plot is the most sensible plot. If one of the variables is discrete and has few values, then the plot will look something like Figure 3.11. If both variables are discrete with lots of possible values, or both variables are continuous, then we will get a traditional scatter plot. For example, we might wish to examine the relationship, if any, between manganese and barium. This scatter plot is shown in Figure 3.14. To make it more interesting, I have replaced the plotting symbols with the bottle numbers. This lets us say that, for example, there appears to be a strong positive linear relationship between the two elements and also that the measurements from bottle number 5 are quite low in both elements.

3.6 Annotating and embellishing plots

There are a number of small things that we can add to plots to make them more informative. Note my choice of words there, however: "...to make them more informative." If our additions do not improve the understanding of our plots, then we should remove them.

3.6.1 Legends

Adding a legend to a plot can be useful when we are displaying information about multiple groups or cases. For example, I added a legend to Figure 3.7 that let the reader see which colors matched which race code. I will now confess that I deliberately rearranged the colors so that the Caucasian group

	Simulated		
Genuine	0	1	2
0	8	1	0
1	3	0	1
2	2	0	0

TABLE 3.2: Number of misleading signatures in comparisons of 16 genuine signatures and 64 simulated signatures from 15 document examiners

was black and African American group was white. The reason I did this was to highlight the usefulness of a legend. One might be tempted to add a legend to Figure 3.14. However, the plotting symbols used are self-explanatory and a legend would be overkill. If you add a legend to a plot, make sure that the legend labels are informative but concise, and that the size of the legend does not lead to its dominating the plot itself. Also, in general it is not a good idea to draw a box around your legend. The box is unnecessary and a distraction to the eye.

3.6.2 Lines and smoothers

Straight lines and smoothing lines can be a good addition to a scatter plot of two quantitative variables. We may add a straight line to a plot to show how well, or not, the data conform to a linear trend. A "smoothing" line may be sensible if the data follow a trend that is not well described by a straight line. Both of these additions, however, can be a distraction.

3.6.2.1 Smoothers

Smoothing lines are called locally weighted regression lines in statistics or sometimes **lowess** or **loess** lines.

The idea behind lowess lines is relatively simple. The procedure is demonstrated in Figure 3.15. In step 1, I have plotted some data that actually comes

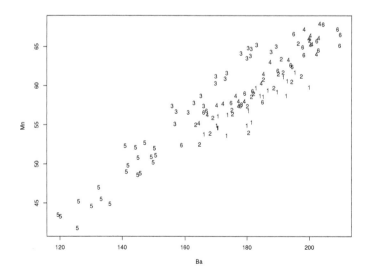

FIGURE 3.14: Concentration of manganese (Mn) versus barium (Ba) in six beer bottles

from a quadratic model. In step 2, we divide the range of the variable on the horizontal axis into equally spaced intervals. In step 3, we fit a straight line (a simple linear regression model) and make a prediction based on that line for the middle of the range. The predicted lines and midpoints are shown by the dashed line and the black dots. Finally, in step 4, we join up our sequence of predictions. In this example I have used straight lines, but in practice a smooth curve called a cubic spline is used.

Smoothers are good for highlighting non-linear trends in the data or for guiding us in the choice of model we might choose for our data. Do be aware, however, that smoothers should not be used when there is a small amount of data ($n < 30$), or when there really is no pattern of interest. I warn against the latter because the observer's eye is naturally drawn to the line regardless of whether it is sensible.

3.6.3 Text and point highlighting

The addition of text to a plot can be helpful in certain circumstances. I will divide the addition into two cases. Firstly, we might consider labeling individual points on a plot. Situations where this might be useful are if we want to highlight a set of extreme points or perhaps a particular group of observations. Figure 3.14 is an example of the latter. In later chapters, when we look at diagnostic plots for linear models, we will see that R automatically labels points that have large residuals or strong influence over the fitted

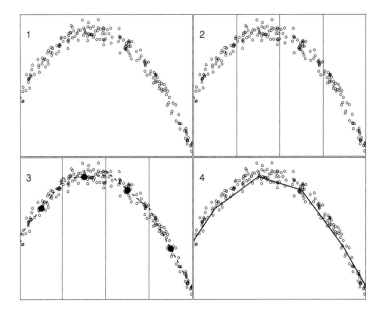

FIGURE 3.15: Fitting a locally weighted regression line or lowess line

model. Secondly, we might want to add text to a plot when we have used a model to describe the data. For example, as I have noted already, it is not uncommon for people to add a simple linear regression line to a scatter plot. We have not discussed regression yet, but that is unimportant. A simple linear regression line is described by an intercept and a slope and the squared multiple correlation coefficient or R^2 is often used as measure of how well the model explains the data. Inclusion of these figures on the plot can be very helpful to the reader.

We may also choose to highlight one or more points on a plot with either a differently shaped plotting symbol or a differently colored plotting symbol (or both). If you do this, take care to make sure that the symbol can be easily distinguished from the rest of the data, both in terms of shape and color. There are no real color examples in this book, but you have seen a number of examples so far where I have increased the size and shape of the plotting symbol so it is easy to understand which point I am talking about.

3.6.4 Color

The addition of color to a plot can be very helpful. For example, when I add fitted lines to a plot, I like to color them red. The reason for this is that red is easily distinguished from the black circles that show the observations. The downside is that this information is often lost when it comes to publication, as the red becomes a gray in print. If you choose to use color, then you should consider the following points:

- Can the colors I use be easily distinguished from each other and from the background? For example, yellow dots on a dark blue background are easy to see, but yellow dots on a white background are not. Similarly, purple and pink plotting symbols are hard to distinguish from each other, whereas red and blue are not.

- Do the colors still make sense if my article or report is printed in grayscale? If not, you might consider changing the shape of your plotting symbols rather than the color.

- Do the colors that I have used on the screen look the same when I print them out? The differences between screen colors and printer colors can sometimes be quite substantial. Furthermore, the paper type and quality can also affect the appearance of the dried ink or toner. It is a good idea to have a palette of the colors you use regularly printed out on the type of paper you intend to use.

3.6.5 Arrows, circles, and everything else

We can add anything we like to a plot. For example, if we are showing a statistical clustering technique, then we might circle a set of points with a circle or an ellipse. Arrows can be used to point to certain features with a text label. There are some specialized plots that use line segments to describe multi-dimensional data. At the end of the day, we must return to our question "Does it help us explain the data or make our point?"

3.7 R graphics tutorial

So far we have seen a lot of theory and advice on how to do things, but we have not learned how to do it. In this section, I will try to give an overview of how to construct simple graphics in R, how to annotate and improve those graphs, and how to get them into your document. Many of the tasks in this section will use data we have discussed previously. Make sure you have installed the **dafs** package and that you type

```
> library(dafs)
```

at the start of each session. This will ensure that the **data** commands in the tasks work.

3.7.1 Drawing bar plots

The basic inputs to the **barplot** function are the heights of the bars. There is no requirement for anything else. However, it is very likely that you will want to label the bars. In this task we will produce a bar plot of all the genotypes present at the Gc locus.

1. Load the FBI-Gc data.

    ```
    > data(gc.df)
    ```

2. Create a table of counts for the genotypes and print it out.

    ```
    > tbl = table(gc.df$genotype)
    > tbl
    
    AA  AB  AC  BB  BC  CC
    29  72 114 136 137 124
    ```

3. Turn the table into a table of proportions by dividing by the table sum.

    ```
    > tbl = tbl/sum(tbl)
    > tbl
    
            AA         AB         AC         BB         BC
    0.04738562 0.11764706 0.18627451 0.22222222 0.22385621
    ```

```
          CC
0.20261438
```

4. Create a bar plot of the proportions like Figure 3.5 with the following features:

    - the y-axis limit going from zero to one
    - the genotypes on the x-axis
    - an x-axis label of *Genotype*
    - a y-axis label of *Proportion*
    - a box around the plot

```
> barplot(tbl, ylim = c(0, 1), xlab = "Genotype",
    ylab = "Proportion")
> box()
```

> **Tip 8: Changing the axis labels**
>
> The x-axis label and y-axis label of any type of plot can be controlled by adding xlab = "x label" or ylab = "y label" to the plot command you are using, where x label and y label are the labels you wish to use. If you want to suppress a particular label, then use xlab = "" or ylab = ""

5. Add the cumulative frequency line to the bar plot.

```
> midpts = barplot(tbl, ylim = c(0, 1), xlab = "Genotype",
    ylab = "Proportion")
> cumppn = cumsum(tbl)
> lines(midpts, cumppn)
> box()
```

Note that when I used the barplot command, I wrote midpts = barplot(.... I did this because the barplot invisibly returns the midpoints of each of the bars. The midpoints are the co-ordinates or positions along the x-axis of the middle of each bar. Many R functions return an invisible result like this.

The lines command takes a set of x values and a set of y values and draws a line from (x_1, y_1) to (x_2, y_2) and so on.

6. Put the proportions (to two decimal places) on the top of each bar.

```
> text(midpts, tbl + 0.05, round(tbl, 2))
```

The `text` command puts text on a plot. I have chosen to plot the text at the x positions given by `midpts` and 5% (0.05) above the row proportions for each bar (`tbl+0.05`).

7. To produce a bar plot for grouped data, we need a table (or a matrix), where there is a row for each group. In this task, we will reproduce Figure 3.7. Firstly, we need the table of genotype counts by race code. We can use the `table` command or the `xtabs` command. The `xtabs` command is useful when the variables of interest are stored in a data frame. It adds a formula interface to the `table` command. I have included both and you will notice later in the book that I will tend to use `xtabs` in preference to `table`.

```
> tbl = table(gc.df$racecode, gc.df$genotype)
> tbl

    AA  AB  AC  BB  BC  CC
  A  1  28   6 117  46   7
  C 19  24  60   4  29  63
  H  9  20  48  15  62  54
```

The code below shows how we could make the same table with `xtabs`. I have included it here for completeness.

```
> tbl1 = xtabs(~racecode + genotype, data = gc.df)
> tbl1

        genotype
racecode AA  AB  AC  BB  BC  CC
       A  1  28   6 117  46   7
       C 19  24  60   4  29  63
       H  9  20  48  15  62  54
```

8. Next, we want to convert the rows of this table into the row proportions. To do this we need the row sums, which can be obtained with the `rowSums` command.

```
> rowSums(tbl)

  A   C   H
205 199 208

> tbl = tbl/rowSums(tbl)
> ## We will just display the table to 4 decimal places
> ## The rounding will have no effect here on the stored
> ## values
> round(tbl,4)
```

```
          AA      AB      AC      BB      BC      CC
A      0.0049  0.1366  0.0293  0.5707  0.2244  0.0341
C      0.0955  0.1206  0.3015  0.0201  0.1457  0.3166
H      0.0433  0.0962  0.2308  0.0721  0.2981  0.2596
```

Those of you who know a bit about matrices know that `tbl/rowSums(tbl)` should not work. However, the general philosophy of R is "try to do the sensible thing," which may not necessarily be the correct thing. In this case, luckily for us, it does do what we want it to do, which is to divide each row in the table by the appropriate row sum. If we were forced to express the code in terms that obey the rules of matrix algebra, we would need to pre-multiply the table by a vector that contained one over each of the row totals.

9. Now we have the information we need to plot the proportions by race code. To make R plot a separate bar for each race code, we add the code `beside = T` to our `barplot` command. If we do not do this, we get a stacked bar plot. We also want to set the colors for each of the race codes: blue for African American, yellow for Caucasian, and red for SW Hispanics. We do this by adding `col = c("blue", "yellow", "red")` to our bar plot command.

```
> barplot(tbl, beside = T, ylim = c(0,1),
          xlab = "Genotype", ylab = "Proportion",
          col = c("blue", "yellow", "red"))
> box()
```

R understands 657 predefined color names, which you can see by typing `colors()`.

Tip 9: Adding color

Most plot commands have a `col =` option. This means that you can specify a color or set of colors that will be used in the plot. How the color is used depends on the plot command. For example, in the `plot` command, R will change the color of the plotting symbols, whereas in the `hist` command R will change the color of the histogram bars.

---

10. Add a legend to the plot, showing how the colors relate to the race code.

```
> legend(1,0.99,
         legend = c("Af. American", "Caucasian",
                    "SW Hispanic"),
         col = "black",
         fill = c("blue", "yellow", "red"), bty = "n")
```

3.7.2 Drawing histograms and kernel density estimates

Histograms and scatter plots are probably the two most frequently used graphs in all of statistics. Accordingly, R makes it very easy to draw a histogram.

1. Load the RI data from Bennett et al. [9].

   > data(bennett.df)

2. The data in bennett.df is stored in a slightly different way than we might imagine. Recall that there are 10 RI measurements for each of 49 panels of glass. These have been stored so that the measurements for each panel are a separate variable in the data frame. For example, the variable X1 contains the 10 RI measurements made on panel 1, the variable X40 contains the 10 RI measurements made on panel 40 and so on. We are not interested in the panel information in this task, so we can use the unlist command to collapse the data into a single vector, which we will call ri.

   > ri = unlist(bennett.df)

3. The hist command produces a histogram.

   > hist(ri)

4. The density command creates a density estimate, but it does not actually draw it. To plot the density estimate, you need to use the plot command.

   > plot(density(ri))

5. It can be good to show both a histogram and a density estimate on the same plot. This lets your readers see where the data actually is, because a histogram does not draw bars where there is no data, and where you estimate the density to be. By default, the hist command uses the counts of the observations that fall into the frequency table "bins" for the vertical scale. We want to add a density estimate, which is on the density scale. Therefore, we need to add prob = T to our hist command to make it plot on the density scale. This will not change the appearance of the plot, just the labels on the y-axis.

   > hist(ri, prob = T)
   > lines(density(ri))

   You will notice that the density estimate curve touches the top of the plot. To overcome, this we need to change the limits of the y-axis. The largest value we need to show is the maximum height of the density estimate. We will want our plot to be a bit bigger than that though, so the density line does not touch the top of the graph. We will make it 10% bigger than it needs to be.

Graphics

```
> d = density(ri)
> yMax = max(d$y) * 1.1
> hist(ri, prob = T, ylim = c(0, yMax))
> lines(d)
```

> **Tip 10: Changing the extent of the axes**
>
> Most plot commands understand the optional arguments xlim and ylim. These arguments let you specify the maximum and minimum values for the x-axis and the y-axis, respectively. To use them, you must supply a vector containing two elements with the first element strictly less than the second. That is, the minimum x- or y-value must be smaller than the maximum x- or y-value. These commands are very useful when you are trying to make sure different plots have the same scale.

6. Finally, we will clean up our plot a little bit and make it nicer. We will:

    - change the x-axis label;
    - remove the graph title;
    - make the histogram a little more fine scale, i.e., use smaller width bars;
    - make the x-axis touch the bottom of the histogram; and
    - make the density estimate line width thicker.

```
> par(xaxs = "i", yaxs = "i")
> d = density(ri)
> yMax = max(d$y)*1.1
> breaks = seq(min(d$x),max(d$x),length = 20)
> hist(ri, prob = T, ylim = c(0, yMax),
        xlab = "Refractive Index",
        main = "", breaks = breaks)
> lines(d, lwd = 2)
> box()
```

> **Tip 11: The par command**
>
> par is a low-level function that controls nearly every aspect of how a plotting device behaves. There is far more complexity in par than I will cover in this book. Therefore, if we use par, you can find out what it is doing by looking at the help file (?par).

3.7.3 Drawing box plots

The `boxplot` command can be used in a number of different ways. We will cover two of them, being:

- The different groups of data are in different columns of the data frame;

- The data are stored so that one column of the data frame represents the data and the other column tells us which group the data belongs to.

The data frame `bennett.df` is an example of the first situation and the data frame `bottle.df` is an example of the second.

1. Load the Bennett data set.

    ```
    > data(bennett.df)
    ```

2. Produce a box plot of the first 10 groups.

    ```
    > boxplot(bennett.df[, 1:10])
    ```

3. Create a box plot of the zirconium concentration of the different bottles like Figure 3.13.

    ```
    > data(bottle.df)
    > boxplot(Zr ~ Number, data = bottle.df)
    ```

4. Add the actual data points to the box plot. Use the plotting symbol "x".

    ```
    > points(bottle.df$Number, bottle.df$Zr, pch = "x")
    ```

Tip 12: Changing plotting symbols and line types

Most plotting commands in R have optional arguments `pch` and . These arguments control the plotting symbol and line type, respectively. The plotting symbol can be a number from 0 to 20 representing any one of 21 predefined symbols, or a single character enclosed in quotes. The line can be a number from 0 to 6 representing the 7 predefined line types (although 0 is blank). More help can be obtained on line types by looking at the section on `lty` in the help file for the `par` command (`?par`).

3.7.4 Drawing scatter plots

The workhorse of the R graphics commands is the `plot` command. Many of the graphs you have seen so far are actually drawn by the `plot` command. In this tutorial section, I will show you some of the simplest ways to use `plot`, but bear in mind that there is significantly more complexity and flexibility available.

1. The simplest way to use `plot` is `plot(x,y)`, where x is a set of x-values and y is a y-values. `plot` will draw a set of points at $(x_1, y_1), (x_2, y_2), \ldots$.

    ```
    > data(bottle.df)
    > plot(bottle.df$Mn, bottle.df$Zr)
    ```

2. There is an alternative syntax that can be quite useful when your data are stored in a data frame. This is called the *formula* syntax because you effectively create an R `formula` object. We will not go too much into that at this point. The basic formula is `y~x` and we read this as "y versus x." Therefore, we can use this syntax to repeat the last task.

    ```
    > plot(Zr ~ Mn, data = bottle.df)
    ```

    Note that we must use the `data =` option to tell R where to find the variables Zr and Mn. Also note that R will not pay any attention to `data =` when you are not using the formula syntax. For example,

    ```
    > plot(Mn, Zr, data = bottle.df)
    ```

    will give you an error.

3. As we noted previously, it can be helpful to add a straight line or a smoothing line to a plot. The function `abline` can be used to add one or more straight lines to a plot. There are a number of ways to use the `abline` function.

    a Adding one or more horizontal or vertical lines. Use the `h =` option for horizontal lines and the `v =` option for vertical lines, e.g.,

    ```
    > plot(Zr~Mn, data = bottle.df)
    ```

    We use `abline` to put "grid" lines on the plot. Remember, though, this is a really bad idea!

    ```
    > abline(v = seq(45,65, by = 5),
             h = seq(60,120, by = 10))
    ```

    b If you know the intercept and slope of the line you want to draw, then you can provide them as a vector to `abline`, e.g.,

    ```
    > plot(Zr~Mn, data = bottle.df)
    > # draw a line with intercept = -52.5 and slope = 2.5
    > abline(c(-52.5,2.5))
    ```

c If you do not know the intercept and the slope, but want them estimated from the data, i.e., you want a simple linear regression line, then you must use the `lm` function. The `lm` function will find the best least squares fit to the data; more on this in later chapters.

```
> fit = lm(Zr~Mn, data = bottle.df)
> ## We will make it red (and thicker) and plot
> ## it on the same plot so we can see how it
> ## compares with my "by eye" line - i.e. much better
> abline(fit, col = "red", lwd = 2)
```

4. To add a smoothing line, we have two possible functions and they both work slightly differently, although they produce similar lines. The command `lowess` takes x- and y-values as its argument and produces a set of x- and y-values that can be used in the `lines` function, e.g.,

```
> plot(Zr ~ Mn, data = bottle.df)
> lines(lowess(bottle.df$Mn, bottle.df$Zr), col = "blue")
```

The `loess` function takes a formula in the same way that `plot` does. However, in order to plot the line, it is necessary either for the x-values to be sorted into order, or for some ordering to take part in the plotting. For example,

```
> # first we create the smoothed line
> loessLine = loess(Zr~Mn, data = bottle.df)
> # then we find out the order of the x values
> o = order(bottle.df$Mn)
> # then we plot the line (segments) in order
> lines(loessLine$x[o], fitted(loessLine)[o], col = "red")
```

5. Occasionally, we need to plot a line and not the individual points. To do this, we can use the `lty = "l"` option of the `plot` command. However, for this to work, the x-values must be in order otherwise your plot will look like it has been attacked by a toddler with a crayon. The reason for this is that when given this option, R considers the x- and y-values as a sequence of points to draw a line between. Therefore, if x_2 is smaller than x_1, the line will go to the left and not to the right. Using the previous task as an example:

```
> loessLine = loess(Zr~Mn, data = bottle.df)
> o = order(bottle.df$Mn)
> x = loessLine$x
> y = fitted(loessLine)
> # without ordering first
> plot(x,y,type = "l")
> # with ordering
> x = x[o]
```

```
> y = y[o]
> plot(x,y,type = "l")
```

3.7.5 Getting your graph out of R and into another program

Producing graphs in R is all very well, but most of us produce a graph because we want to show something in a presentation, a research article, or a book. R has a number of mechanisms for dealing with this, either through the context menu (right click over the graph window) or through commands.

In the section that follows, I am going to provide some information of graphic file formats. You can skip this if you want. It is a justification for my recommendations.

If you are planning to take an R graphic and put it into Microsoft® Word® or Microsoft® PowerPoint®, the simplest thing you can do is right click on the graph after you have drawn it. This will bring up the context menu shown in Figure 3.16. Select "Copy as metafile" from this menu and then paste the image into Word® or PowerPoint®. Note that in PowerPoint® the graph will by default have the same background color as the slide. To change this left click on the image to select it, then right click on the image, and select "Format image..." and change the "Fill" properties to white (or alternatively you can recolor the graph in PowerPoint® so that the lines and points are in high contrast with the background).

If you want to save your graph so that it can be inserted into Word® or PowerPoint®, then I recommend you use the "Save as metafile..." option. Word® can deal with PostScript but it deals with a Windows® metafile (.wmf file) better. It will also allow you to do things like recolor the graphs.

If you are planning to use the graph in LATEX, then select "Save as postscript..."

FIGURE 3.16: The context menu from the R plotting window

from the context menu. If you are using pdfLATEX, then you will need to use a program like `ps2pdf` or the commercial version of Adobe® Acrobat® to change your PostScript file into a PDF file.

3.7.5.1 Bitmap and vector graphic file formats

In the world of computer graphics and electronic publishing, graphic file formats can be divided into two camps, which I will label **bitmap** and **vector**.

A bitmap file is effectively a representation of what you see on your screen. The file consists of a grid of pixels. Each pixel will have a color value associated with it. If the image is black and white, then you can think of the pixels as being on or off, hence the name "bit." When you save your graph in a bitmap format, R decides firstly how many pixels will be in the grid (you can change this), and secondly which pixels are on or off. The appearance of the graphic, therefore, is linked directly to this set of pixels. The consequence of this for you, as a user, is that if you attempt to rescale (i.e., shrink or enlarge) the image or change the aspect ratio to something other than the original, then you will introduce many artifacts in to your image. As an example, Figure 3.17 is a circle which was drawn on a grid of 50×50 pixels and has been scaled to 200% of its original size.

Unsurprisingly, Figure 3.17 has a bad case of the "jaggies." Bitmap file formats include bitmap (bmp), graphic interchange format (gif), joint photographic experts' format (jpg or jpeg), portable network graphics (png), and tagged image file format (tiff). R supports all of these file formats except for gif.

A vector graphic, rather than being described by a set of pixels, is defined by a co-ordinate system and set of graphic or plotting "primitives." Plotting primitives are the basic set of tools we need to draw a picture, such as points, lines, polygons (sometimes called poly paths), and text. A vector graphic file consists of a sequence of instructions telling the computer how the picture is drawn. So, for example, it might say "This graphic has co-ordinates that run from -10 to 10 in the horizontal direction, and -100 to 100 in the vertical direction. Draw a line from the point $(-5, 5)$ to the point $(21,33)$. Draw a circle at point $(0,0)$ of radius 10..." The advantage of storing a graph like this is that changes in scale will not lead to the introduction of the artifacts we have seen in bitmaps. This is not to say that they do not exist, because ultimately how the graphic looks depends on the plotting device. Some common plotting devices are the computer screen and the printer. When a vector graphic is displayed on the screen, the computer works out a translation, or mapping, between the co-ordinate system of the vector graphic and the bitmap that is the screen. Similarly, when it is printed on paper, the resolution (how many dots per inch or dpi) of the printer dictates the mapping, and how smooth the image is. Vector graphic file formats include PostScript (ps) and encapsulated PostScript (eps), the portable document format (pdf), Windows metafile (wmf) and enhanced Windows metafile (emf), and scalable vector

Graphics

graphics (svg). R directly supports all of these formats except svg. Note that the difference between encapsulated PostScript and PostScript is mostly historic. R's PostScript driver produces encapsulated PostScript. This is true also for wmf and emf files.

3.7.5.2 Using R commands to save graphs

R has a command for every graphic file format it supports that allows a user to save files to disk with R code, rather than using the menus. This is handy to know if you have a large amount of R code, with lots of pictures to be drawn and saved. For example, when I write an article, subtle changes to data may mean that six to seven pictures need to be changed in response. To have to save each of these by hand quickly becomes irritating; therefore, I use R commands to do this automatically. The R commands for each graphics device are given in Table 3.3. The basic syntax is

```
> plot.device("filename")
> plot commands ...
> graphics.off()
```

FIGURE 3.17: A bitmap circle does not scale well

Device/format	R Command
Bitmap	`bmp`
JPEG	`jpeg`
PDF	`pdf`
PostScript	`postscript`
PNG	`png`
TIFF	`tiff`
Windows metafile	`win.metafile`
Windows printer	`win.print`
Windows screen	`windows`

TABLE 3.3: R commands for various plotting devices/formats

where `plot.device` is one of the commands from Table 3.3 and `filename` is the name of the file you wish to save your graph to. The command `graphics.off()` tells R that you have finished drawing and to save the file to disk. For example, to produce a PDF of Figure 3.13 called `ZrBoxplot.pdf`, my code would be

```
> pdf("ZrBoxplot.pdf")
> data(bottle.df)
> boxplot(Zr ~ Number, data = bottle.df)
> graphics.off()
```

Tip 13: Multi-page PDFs and metafiles

One of the cool things about the `pdf` and `win.metafile` functions is that if you do not call `graphics.off` in between successive calls to `plot`, then they will produce a multi-page file. This means if you just want to send a set of plots to someone, you can do it in a single PDF. The different plots will appear as different pages in the PDF file.

3.8 Further reading

My colleague, Paul Murrell, is considered to be the leading expert in R graphics. He is a member of R Core who are the twenty or so core maintainers of the R project. Paul's book, *R Graphics* [14], therefore, is essential reading for people who want to get the most out of R's graphics system. Bear in mind that this book is not a "How To" book. It has a hundred-plus pages on a comprehensive introduction to R's graphics, and then several more hundred pages on more advanced topics. It is a very good reference if you need to know how R does things and how to change them. Deepayan Sarkar, another member of R Core, is the author of an R package called `lattice` and a book on the package, *Lattice: Multivariate Data Visualization with R* [15]. `lattice` is good for investigating multivariate data sets or data with a lot of stucture. Hadley Wickham is a former student of Paul Murrell. He has a relatively new book, *ggplot2: Elegant Graphics for Data Analysis* which describes the R package `ggplot2`. This package is for those of you who want to be able to do *everything and anything* with R's graphics system.

Chapter 4

Hypothesis tests and sampling theory

The occasions ... in which quantitative data are collected solely with the object of proving or disproving a given hypothesis are relatively rare – F. Yates, Statistician.

4.1 Who should read this chapter?

This chapter is primarily aimed at readers who have never had any formal statistical training at all. If you feel comfortable with the ideas of the sampling distribution of sample statistics, hypothesis tests and confidence intervals you can probably skip this. However, if it has been a while since you last studied statistics you might like to flip through this chapter as a refresher. Once again I will aim to keep the formulas to a minimum and concentrate on the core ideas and practice.

4.2 Topics covered in this chapter

In this chapter we shall cover the absolute essentials of statistical hypothesis tests and confidence intervals and some of the theory behind them. Why, you might ask? In the chapters that follow, we will need to be able to understand hypothesis tests and confidence intervals so that we can decide whether the results we see from a statistical method are meaningful or significant. To do this, we need to understand the various bits and pieces that go in the output. We will also cover the parts of R that let you do things like carry out a two sample t-test, calculate a confidence interval, and find a P-value.

4.3 Additional reading

There is no way that we can cover all of the detail of a standard first year statistics course in one chapter. If you would like additional information, then I would recommend Seber and Wild"s [16] *Chance Encounters* or Moore and McCabe's [17] *Introduction to the Practice of Statistics.* Both of these are good introductory texts, and I have successfully used them in large first year or freshman university classes. The biologists amongst my audience may also find Sokal and Rohlf's [18] *Biometry* a useful text as it is extremely popular in the biological sciences.

4.4 Statistical distributions

As I sit down to write this section, I can hear a collective groan from those of you who have done statistics before. I am going to try and overcome the prejudices you might have, which are undoubtedly caused by many tedious, repetitive calculations that involved the manipulation of quantities so that statistical tables could be used. I will promise you now we will not use statistical tables at all in this book. If you can get to a copy of Excel® or R, then that is all you will need.

The reason we need this section is that many of the tests and techniques rely on certain assumptions. The assumptions often are that the data, or a statistic calculated from the data, follow a specific statistical distribution. Sometimes these are reasonable assumptions to make, and sometimes they are not. However, if we do not know what the assumptions mean, then we cannot judge whether they are sensible.

I will also introduce the distributions in order of use rather than taking the standard statistics approach of introducing discrete distributions then continuous distributions.

4.4.1 Some concepts and notation

In this chapter, there are a number of terms that I will use, and so it is necessary that we define them.

A **random experiment** is an experiment where we cannot predict the outcome in advance. We might know the range of possible outcomes, but we do not know the specific outcome until we carry out the experiment. Experiment is used here in the loosest sense. Some forensic examples of random experiments might be:

Hypothesis tests and sampling theory

- Counting the number of fibers found on a car seat;
- Measuring the length of a footprint;
- Determining the RI of a piece of glass;
- Measuring the spectra of a paint sample.

A **random variable** is a variable that measures the outcome of a random experiment. Some examples of random variables, using the random experiments in the previous paragraph, are

- The number of fibers;
- The length in millimeters (mm);
- The RI;
- The positions or wavelengths at which peaks occurred.

It is statistical convention to denote a random variable with an uppercase italic letter, and the outcome of a specific random experiment with a lowercase italic letter, e.g., X is a random variable and an outcome from a particular measurement, x. This notation lets us talk about the probability of an outcome or an event like $X = x$. It also lets us talk about the properties of the sample mean. The sample mean, $\overline{X}$, is a random variable. $\overline{x}$ is an outcome of the sample mean for a particular data set.

Many people think they understand **probability** when very clearly they do not. At its simplest, a probability is a number that varies between 0 and 1 that lets us express the chance that an event or an outcome from a random experiment will happen. A probability of 0 means that the event will never occur and a probability of 1 means that the event will always occur. Some people choose to express probabilities as percentages, so rather than say the probability is 0.1, they say the probability is 10%. This is completely acceptable and does not cloud our understanding. Probabilities are also sometimes expressed in the form of odds. This can be useful, but it does introduce confusion because the transformation between probability and **odds** is not linear. For example, if I tell you that the probability of a certain event occurring is $P = 0.25$ (25%), then the odds are

$$Odds = \frac{P}{1-P} = \frac{0.25}{0.75} = 1/3 \tag{4.1}$$

Odds are usually expressed as a ratio so our probability of 0.25 becomes odds of 1:3 or "one to three against." The confusion arises because people read the ratio 1:3 as a fraction 1/3 and then believe that the probability is also $1/3 \approx 0.33$, which is clearly incorrect. The formula for converting odds back to a probability is

$$P = \frac{Odds}{1 + Odds} \tag{4.2}$$

Probabilities may be derived from theoretical models. For example, when someone tells you that the probability of getting a head or a tail from a toss of a fair coin is one-half, they (and you) are assuming a certain idealization of a coin. In reality, for most coins, the probability will be close to a half, but not exactly. If you do not believe me, then you might consider Rai stones, which were the currency of the Yapese people in Micronesia. Rai stones are circular disks carved out of limestone with a large hole in the middle. The size of the stones varies widely with the largest being 3 meters (10 ft) in diameter, 0.5 meters (1.5 ft) thick and weighing 4 metric tons (8,800 lb) [19]. It may be difficult to toss one of these coins fairly.

Probabilities may also be derived from long-run averages or large sample frequencies. These are the probabilities that we are perhaps most familiar with. For example, we do not know the probability of observing an A allele at the Gc locus in a randomly chosen member of the South Western Hispanic population of the United States, but we can estimate it by collecting a sample of people and working out its frequency. Similarly, we do not know the probability of observing a fragment of glass with RI 1.518, but we may estimate it from non-matching case work glass.

Probabilities may also come from someone's "best guess." We call these **subjective probabilities** because they are influenced by an individual's beliefs. There is a long tradition in statistics of saying that subjectivity is bad. However, we also have a tradition in forensic science of respecting the opinions of experts. An expert in a particular field may be able to provide a more accurate assessment of a probability than a "layman." We can use this information as long as we accept that it is an opinion and that there is some uncertainty associated with it.

4.4.2 The normal distribution

The normal or Gaussian distribution is probably the most "famous" distribution in statistics. It is an extremely useful distribution because there are many practical instances of "normal" or "normal-like" behavior. The normal distribution is a mathematical formalization of the idea that "observations tend to cluster around the mean." This is something that we have all probably observed in one circumstance or another. However, there is a situation where this observation becomes stronger and that is in the behavior of the sample mean. The behavior of the sample mean is described by the Central Limit Theorem. What it means is that in most situations we can imagine, we can rely on the approximate normality of the sample mean, which turns out to be a very useful thing. Why? Firstly, we know a lot about the behavior of the normal distribution. Secondly, a large number of the techniques rely on normality.

Data are said to be **normally distributed** if they "follow a normal distribution." That definition is close to useless. In practice, we say data are normally distributed if a histogram or a kernel density estimate of the data is

- unimodal;

- symmetric;

- and the observations tend to cluster around the mean.

The normal distribution is characterized by the mean μ and the standard deviation σ. The mean specifies where the distribution is centered and the standard deviation specifies the spread of the distribution. Figure 4.1 shows normal distributions with different means. In this plot $\mu_1 < \mu_2 < \mu_3$. Figure 4.2 shows normal distributions with different standard deviations. In this plot $\sigma_1 < \sigma_2 < \sigma_3$. When $\mu = 0$ and $\sigma = 1$, the normal distribution is referred to as the **standard normal**. This is mostly a historical irrelevancy with the exception of one small fact. The process of **standardizing** an observation or data refers to the mathematical process of "subtracting the mean and dividing by the standard deviation." That is, if you have data $x_1, x_2, \ldots, x_n$ with mean $\overline{x}$ and standard deviation s, then the standardized observations are

$$z_i = \frac{x_i - \overline{x}}{s} \tag{4.3}$$

When we do this, we call the resulting numbers the z-**scores**. The z-scores tell us how many standard deviations each observation is away from the mean. For example, the mean of the RIs in the Bennett data set is 1.51999 and the

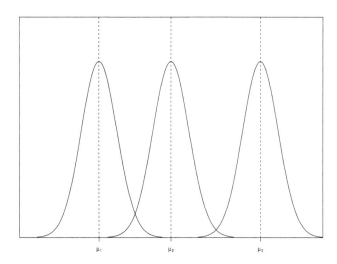

FIGURE 4.1: Normal distributions with the same standard deviation and different means

standard deviation is 4.175×10^{-5}. The largest observation is 1.52016. This observation has a z-score of

$$z = \frac{1.52016 - 1.51999}{4.175 \times 10^{-5}} = 4.1$$

This means that the maximum is 4.1 standard deviations above the mean. We can use this information in conjunction with what is sometimes known as the **3-σ rule** or the **empirical rule**. If a random variable (or data) is normally distributed, then

- approximately 68% of the observations lie within one standard deviation of the mean;

- approximately 95% of the observations lie within two standard deviations of the mean;

- and approximately 99.7% of the observations lie within three standard deviations of the mean.

Therefore, in our example, our observation has a z-score of 4.1, which is exceptionally high. We could conclude from this that

a. the data are not normally distributed; or

b. the data point is an outlier; or

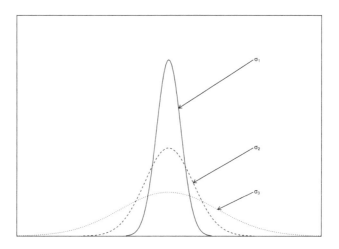

FIGURE 4.2: Normal distributions with the same mean but different standard deviations

c. the data consist of a mixture of normal distributions.

Bennett et al. [9] favored the last explanation, and this was later backed up by the work of Newton et al. [12].

4.4.3 Student's t-distribution

Student's t-distribution [20] is closely related to the normal distribution in that it is used to model the behavior of observations that cluster around the mean. The principal difference is that t-distribution has "fatter" tails that allow more variability in the data than the normal distributions. It is primarily used in situations where we have a small amount of data and therefore increased uncertainty in our estimates of the mean and the standard deviation. The t-distribution allows for this extra uncertainty. The t-distribution is symmetric around zero, and its behavior is controlled by a parameter called the **degrees of freedom**. Figure 4.3 shows the effect of the degrees of freedom. As the degrees of freedom decrease, the "peak" of the curve gets lower, and the tails get fatter. The degrees of freedom (usually) relate to the sample size, but they are determined on a case by case basis. We will not worry too much about these formulas as R can work out the degrees of freedom in most situations. Note that "Student" is capitalized because it is the pen name of William Sealy Gosset who discovered it.

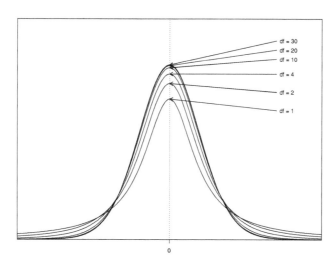

FIGURE 4.3: Student's t-distribution gets a lower peak and fatter tails as the degrees of freedom decrease

4.4.4 The binomial distribution

The **binomial distribution** is used when we have data that record "the number of successes in a fixed number of experiments." Some examples where binomial data might arise are

- A sample of 100 unrelated people is genotyped and the number of A alleles at the Gc locus are counted for each person;
- A particular DNA profile is genotyped 10 times and the number of "dropout" events are counted;
- A random sample of 50 adult males is taken and the number of men wearing size 12 shoes or larger is recorded.

There are four key assumptions that must be met in order for the binomial distribution to be used. These are

1. In each trial there must only be two possible outcomes: "success" and "failure";
2. Each trial must be independent of all previous trials;
3. Each trial must have a constant probability of success;
4. The number of trials must be fixed in advance.

Let us take our third example to see how we might apply these assumptions. A trial in this example is seeing whether a particular male has size 12 shoes or bigger. Therefore, each trial has only two possible outcomes: success = "size 12 or bigger," and failure = "smaller than size 12." Each trial is independent of the previous trial because we have no real reason to think that the shoe size of one man should be related to the shoe size of another man. We might come up with spurious arguments like "What if they are brothers?" However, firstly the description experiment specified the sample was random, and therefore two brothers are unlikely to be in the same small sample. Secondly, even if there are minor violations to this assumption, it is unlikely to affect the overall behavior of our statistics or tests in the long run. The probability of success is constant in this trial because we have a belief that there is a constant population probability of "size 12 or larger." Finally, we said that the sample size is 50, so the number of trials has been fixed at 50.

The behavior binomial distribution is controlled by two parameters, n, the number of trials, and p, the probability of success. We routinely (and usually unknowingly) use the formula for the binomial mean, which is np. For example, if we tell someone that the population frequency of red hair is 7% (0.07) and we have taken a sample of 1,000, then many people will work out that there should be about $1,000 \times 0.07 = 70$ redheads in our sample.

The binomial distribution a key component of the **logistic regression** model that we will encounter in Chapter 6. The core idea is that if we have binomial data where we believe the probability of success is affected by another variable, then we can use logistic regression to model this data.

4.4.5 The Poisson distribution

The **Poisson** distribution is very useful for dealing with count data where there is no fixed number of trials. The name of the distribution is capitalized because it was discovered by Siméon-Denis Poisson (1781–1840) and published, together with his probability theory, in 1838 in his work *Recherches sur la probabilité des jugements en matière criminelle et matière civile* [21]. The Poisson distribution has a single parameter, its mean λ. It is useful in modeling situations where the variable being measured may take on values 0, 1, 2, ... For example, in a later chapter we will look at some experiments where we model the number of glass fragments that have fallen to the ground after it is broken. The number of fibers recovered in different tapings could also be modeled with the Poisson distribution.

4.4.6 The χ^2-distribution

The χ^2 (or chi-squared) distribution is encountered mainly in two situations that are actually related. Tests of **goodness-of-fit** and tests of **independence** both use the χ^2-distribution. The χ^2-distribution is characterized by a degrees of freedom parameter. Like the t-distribution, the degrees of freedom parameter is related to the sample size.

4.4.7 The F-distribution

We will routinely use the F-distribution in the interpretation of linear models. The F-distribution has no natural interpretation but is useful when the statistic of interest is a ratio of variances.

4.4.8 Distribution terminology

If we say that data come from a particular (theoretical) distribution, then we mean that if we collected enough data, the histogram or kernel density would look like the curve that describes the distribution. If the distribution is discrete, like the binomial or Poisson, for example, then this curve is called a **probability function** and can be represented as a histogram. The areas of the bars represent the probability of the outcomes. If the distribution is continuous, like the normal or t, then the curve is called a **probability density function** or **pdf**. Defining these terms lets us say that if data follow or come from a particular distribution then we expect the empirical pdf will look very similar to the theoretical pdf. The probability of a particular value for a discrete distribution or for a range of values for a continuous distribution is given by another curve called the **cumulative distribution function** or **cdf**. If data follow or come from a particular distribution, then the empirical cdf will be similar to the theoretical cdf for that distribution.

4.5 Introduction to statistical hypothesis testing

Statistical hypothesis testing is a formal, objective, repeatable way of making inferential statements based on data. That is, it is a scientific method for assessing the importance or relevance of experimental findings based on one or more sets of data. Readers who are familiar with my published work may recall that I have been critical of hypothesis testing in the past. This is true, but the context is different. In this book, I am interested in the use of statistics primarily for experimentation and research. I am not talking about the interpretation of evidence or the presentation of evidence in court. In those situations Bayesian methodology is usually more appropriate. There are Bayesian versions of everything I present in this chapter and probably the rest of this book. However, I think in order to use those Bayesian methods it is necessary to understand the standard or classical frequentist methodology first. There are some limits to the inferential statements that we can make based on the techniques that I will talk about, but as long as we understand the statements we make, then those limitations become less relevant.

4.5.1 Statistical inference

Statistical inference is the process of using a "sample" of data to say something about a "population." I have put those words in quotation marks because their meaning depends on context. Perhaps a more formal statement is: "Statistical inference is the process of using sample statistics to make statements about an unknown population parameter." A **parameter** is just another word for a statistic. For example, in the previous chapter, we observed a difference between the frequency of the *AB* genotype for African Americans and Caucasians (0.137 vs. 0.121). These frequencies are both estimates of the population probabilities of the *AB* genotype in the respective populations. Another example is the mean RI in the Bennett data set. This is a sample mean and it is an estimate of the overall mean RI of the window.

4.5.1.1 Notation

A convention in statistics is to represent parameters or population values with Greek letters and a sample estimate as a letter from the alphabet or Greek letter with a "hat." For example, we represent the population standard deviation with the Greek letter σ, and the sample standard deviation with the letter s or sometimes as $\hat{\sigma}$. Similarly, we represent population probabilities as π and a sample estimate (or sample frequency) as p or $\hat{\pi}$.[1]

---

[1] This example can be a little confusing, however, as people sometimes use p and $\hat{p}$. The intent should be obvious from context.

4.5.2 A general framework for hypothesis tests

The steps involved in a hypothesis test are generally the same regardless of the type of test being carried out.

1. Ask a question;

2. Formulate your question statistically—that is, find a statistic you think might answer your question;

3. Propose a null hypothesis;

4. Propose an alternative hypothesis;

5. Calculate your test statistic;

6. Calculate the P-value;

7. Interpret the P-value.

Clearly, there are some terms here that we have not defined. The **null hypothesis** might be called "the hypothesis of no interest." In general, the null hypothesis is one of no effect or no difference. For example, if we are interested in the difference between the two population probabilities, then the null hypothesis is that there is no difference. If we are interested in the difference between group means, then the null hypothesis will be that there is no difference between the groups. Corresponding to the null hypothesis is the **alternative hypothesis**. As you might imagine, this is generally the hypothesis we are interested in because it is a hypothesis of an effect. Note that the null and alternative hypotheses do not have to be mutually exhaustive. We will discuss this point when it is more appropriate. The appropriate **test statistic** depends on the situation. The keys to choosing an appropriate test statistic are

- It is a good summary of what you are interested in;

- It has different behavior under each of the hypotheses. That is, it behaves differently if the alternative hypothesis is true rather than the null hypothesis;

- The distribution of the statistic when the null hypothesis is true is known.

The last point is flexible, but if the statistic conforms to a well-known statistical distribution, then we can use our knowledge of that distribution to calculate the P-value easily.

> The **P-value** is the probability of observing a test statistic as large as or larger than the one you have observed **if the null hypothesis is true**.

Understanding the definition of the P-value given above is vital to the interpretation of the P-value. Note especially that the definition does not state in any way that the P-value has anything to do with the truth or falsity of the null hypothesis.

Tip 14: P-values

P-values **do not** tell us the probability that the null hypothesis is true. P-values **do not** tell us the probability that the alternative hypothesis is true. P-values tell us the probability of observing our test statistic by random chance alone (i.e., **if** the null hypothesis is true).

In simplistic terms, however, we might regard the P-value as a measure of evidence against the null the hypothesis. Please be aware that this is not strictly true and should you be accosted by fanatical Bayesians you are going to have concede the point. However, this is not a religious book. In general, the smaller the P-value, the stronger we regard the evidence against the null hypothesis. The reasoning is easier to give with reference to Figure 4.4. Figure 4.4 shows the distribution of some test statistic that I will call T under both the null hypothesis, H_0, and the alternative, H_1. That is, if the null hypothesis is true then the probability of the different values that T may

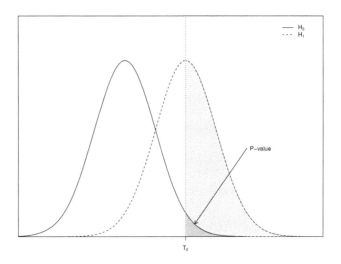

FIGURE 4.4: A visual representation of a P-value

take is described by (the area under) the curve with a solid line. If, on the other hand, the alternative is true, then the probability of the different values that T may take is described by (the area under) the curve with a dashed line. Now assume that we have taken a sample and calculated a test statistic, which I will call T_0. T_0 is shown as the dotted vertical line in Figure 4.4. If H_0 is true, then the dark gray shaded area is the P-value. If H_1 is true, the probability of seeing a test statistic as big as T_0 or bigger is the light gray shaded area *and* the dark gray shaded area. Clearly, this probability is much larger than the P-value. Hopefully, you can see that if P-value is small, then the test statistic is unlikely under the null hypothesis. The implication or inference that people draw, therefore, is that the test statistic must be very likely under the alternative. In fact we do not know[2] if this is true, but in many situations, it is a reasonable assumption.

Tip 15: Interpreting P-values

If the P-value is

- greater than 0.1, then we say, "there is **no evidence** against H_0"

- between 0.05 and 0.1, then we say, "there is **weak evidence** against H_0

- between 0.01 and 0.05, then we say, "there is **evidence** against H_0"

- between 0.001 and 0.01, then we say, "there is **strong evidence** against H_0"

- less than 0.001, then we say, "there is **very strong evidence** against H_0"

4.5.3 Confidence intervals

We use confidence intervals to quantify the uncertainty in our estimates. Confidence intervals are expressed as a range with an associated **confidence level**. For example, we might say, "Our estimate of the mean foot length of men who wear size 12 shoes is 305 mm. A 95% confidence interval for this estimate is 297 mm to 313 mm." The confidence level (e.g. 95%) is a statement about the random properties of the confidence interval and cannot be interpreted as a probability. There is more on this in Section 4.5.4.

---

[2]This is the essence of the Bayesian criticism.

4.5.3.1 The relationship between hypothesis tests and confidence intervals

There is a relationship between **two-tailed** hypothesis tests and confidence intervals. A two-tailed hypothesis test is one where there is no direction in the alternative hypothesis. This is usually the case when the alternative takes the form
$$H_1 : \theta \neq \theta_0$$
where θ is the parameter of interest, and θ_0 is the hypothesized value. For example,
$$H_1 : \pi \neq 0.5$$
and
$$H_1 : \mu_1 - \mu_2 \neq 0$$
In the first example, our alternative hypothesis says π may be any value but 0.5. Evidence of support for this hypothesis would be if the sample proportion $\hat{\pi}$ was either substantially smaller than 0.5, or substatially greater than 0.5. In the second example, our alternative hypothesis says that there a difference between the means. We do not care whether the difference is positive or negative, just whether there is a difference. Our P-value is calculated by adding the lower and upper tail probability together. That is, if our test statistic, T, has value T_0 and we assume $T_0 > 0$, then the P-value is
$$P = \Pr(T \leq T_0) + \Pr(T \geq T_0)$$
This second example illustrates the "two-tail" part of the hypothesis. That is, we consider the probability in both tails of the distribution.

A confidence interval and a two-tailed hypothesis convey some of the same information. If a $100 \times (1-\alpha)\%$ confidence interval contains the hypothesized value θ_0, then we would fail to reject the null hypothesis at the α level of significance. If the interval does not contain the hypothesized value, then we would reject the null hypothesis at the α level of significance.

For example, assume we are making an inference about a population proportion, π and our hypotheses are
$$H_0 : \pi = 0.5$$
$$H_1 : \pi \neq 0.5$$

We collect some data and observe a sample proportion of $\hat{\pi} = 0.56$. We calculate a 95% confidence interval for π, which is [0.48, 0.62]. This interval contains the hypothesized value, 0.5, so we would say that we have insufficient evidence to reject the null hypothesis at the 0.05 level. If, on the other hand, the interval was [0.54, 0.58], then we would reject the null hypothesis at the 0.05 level.

This relationship is used almost routinely in later chapters in this book.

Typically, the hypothesized value is zero. Therefore, statements such as "The (confidence) interval does not contain zero, therefore the coefficient is significant" indicate that we are using the confidence interval to test the hypothesis $\theta = 0$, where θ is the parameter of interest.

4.5.4 Statistically significant, significance level, significantly different, confidence, and other confusing phrases

Some readers will have seen phrases like "this difference is significant at the 5% level," "we have used a 0.01 level of significance," or "we are 95% confident that there is a significant difference." Statistical significance is related to P-values and Type I error. A **Type I error** is the error of declaring a difference between two groups when in fact there is none. If a test is carried out with a significance level of 0.05, then this is a statement that the experimenter accepts they will (on average) falsely declare there is a difference about 5% = 100×0.05% of the time. The first source of confusion is that people often express significance levels as either proportions or percentages. That is, one might use a 0.05 significance level or a 5% significance level. Secondly, it is also common to use $1 - \alpha$ or $100(1-\alpha)$%, where α is a number between 0 and 1, for example, "at the 0.95 level" or the "95% level." In all of these cases, the relevant quantity is α. If the P-value falls below α, then the experimenter will declare a difference or an effect to be **significant** or **statistically significant**. Often in scientific literature, the significance level is implicitly set at 0.05. **There is no scientific basis for this value.** The values 0.05 and 0.01 are popular because these are the values chosen by Sir Ronald Aylmer Fisher [22], the "father" of modern statistics. In choosing these values, the experimenter accepts that they will make a Type I error approximately one time in 20 for 0.05 and one time in 100 for 0.01. They do, however, seem to work well in practice.

Adding further to the confusion is the concept of a confidence level. Confidence levels are usually associated with **confidence intervals**. The confidence level is usually expressed as a percentage, e.g., a 95% confidence interval. The confidence level dictates the width of the interval. For example, a 95% confidence interval will be *wider* than a 90% confidence interval. By wider, I mean that the distance between the lower limit of the interval and the upper limit of the interval will be larger. The confidence level refers to the sampling properties of confidence intervals in general. If a researcher or an experimenter says, "I am 90% confident that the true value lies within this interval," then what they are technically saying is, "If I could take an infinite number of samples of the same size as my study and calculate a confidence interval for each one of them, then approximately 90% of them would capture the true value." The confidence level is a statement about the random nature of confidence intervals in general rather than the specific confidence interval you have calculated. Most people find this extraordinarily confusing or are unaware and choose incorrectly to interpret the confidence interval as a probability, e.g., "there is a 90% chance that the true value is in my interval." This is another source of complaint for Bayesians. A better way of looking at a confidence interval is as an imperfect quantification of the potential uncertainty in an estimate. It is imperfect in the sense that perhaps we cannot interpret it in the way we want. However, it is at least an admission that the numbers we have calculated do have uncertainty associated with them.

In the sections that follow, I will describe two elementary (but useful) statistical tests. These are the two sample t-test, and the χ^2-test of independence. There are many named hypothesis tests, and I could spend the rest of the book describing them. I have limited myself to these two tests as I believe they are the most useful.

4.5.5 The two sample t-test

The **two sample t-test** is used when we have two **independent** groups of subjects or objects for which we have measured a single quantitative variable. We have observed a difference in the sample means of each of these groups and we wish to test the hypothesis that "this difference is real." In reality what we are doing is calculating the chance that the observed difference could have occurred simply from random variability. We will consider two examples.

4.5.5.1 Example 4.1—Differences in RI of different glass strata

Recall that the data of Newton et al. [12] has refractive index measurements from five different strata through the thickness of a piece of float glass. We saw in Figure 3.11 that there appeared to be an appreciable difference between the mean RI for fragments from the float surface stratum and the mean RI for the fragments from the bulk glass.

Figure 4.5 shows the box plot for just these two groups. We can easily see that the difference between the medians for the two strata is large. Similarly, we can see, even accounting for spread, that there is no overlap between the two groups of data at all. Therefore, a two sample t-test using this data should really just confirm our belief that the difference between the means is unlikely to have come from random variation in the data alone. The null hypothesis for the two sample t-test is that there the population means of the two groups are equal. We will test this hypothesis using our sample data. The alternative hypothesis is simply that the means are not equal. We can express this formally as

$$H_0 : \mu_B = \mu_{FS}$$
$$H_1 : \mu_B \neq \mu_{FS}$$

However, it is more usual to express the hypotheses in terms of differences.

That is,

$$H_0 : \mu_B - \mu_{FS} = 0$$
$$H_1 : \mu_B - \mu_{FS} \neq 0$$

These two statements of the hypotheses are equivalent. We use the difference in the sample means to assess the evidence for or against the null hypothesis. However, we must account for the variability present in the samples as well. This is what the test statistic for the two sample t-test does. This is not a book about formulas. Therefore, I will not derive the test statistic or its distribution. I will instead concentrate on the interpretation of the output.

```
        Welch Two Sample t-test

data:  ri by stratum
t = 9.3802, df = 29.019, p-value = 2.744e-10
alternative hypothesis: true difference in means is not equal to 0
95 percent confidence interval:
 0.001652822 0.002574511
sample estimates:
mean in group FS  mean in group B
        1.521270         1.519157
```

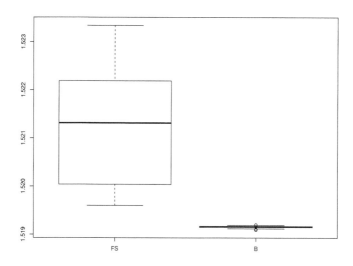

FIGURE 4.5: Box plot of RI measurements for the float surface (FS) and bulk (B) glass

The output above is from R's `t.test` command using the data from the float surface and bulk groups. The `t.test` command is capable of performing one sample, two sample and paired t-tests. The first thing we look at is the P-value, which is $2.744e-10 = 2.74 \times 10^{-10}$. This is an extremely small number; therefore, we can say that there appears to be very strong evidence against the null hypothesis. That is, we believe that there is a difference between the mean RI of the float surface and the bulk glass. The second thing we can look at is the confidence interval. We will formally define the confidence interval later on. Here, we can use the confidence interval to quantify the magnitude of the difference between the two means. In this case, we can say "with 95% confidence the true difference in mean RI lies between 0.0026 and 0.0017." You probably have noticed that I have changed the signs of the confidence and reversed the order. The reason is that it is utterly arbitrary whether we consider the difference $\mu_{FS} - \mu_B$ or $\mu_B - \mu_{FS}$. What is important is the magnitude of the difference. Finally, we might look at the test statistic, which is labeled `t` in the output. The value 9.4 can be read as like a z-score. It says (loosely) that the observed difference is 9.4 **standard deviations** above the mean (of zero). The interpretation is loose because it requires discussion about sampling distributions of test statistics, which I will cover in a later section.

4.5.5.2 Example 4.2—Difference in RI between bulk and near-float surface glass

As a second example, we will examine the difference between the bulk glass and the near float surface. Figure 4.6 shows the box plot for just these two groups. We can see from Figure 4.6 that although there is a difference between the medians that difference might be explained simply by random variation.

```
        Welch Two Sample t-test

data:  ri by stratum
t = 2.0875, df = 52.129, p-value = 0.04175
alternative hypothesis: true difference in means is not equal to 0
95 percent confidence interval:
 5.688924e-07 2.876444e-05
sample estimates:
  mean in group B mean in group NFS
         1.519157          1.519142
```

A two sample t-test using this data shows that there is evidence against the null hypothesis. We make this interpretation because the P-value is 0.042, which is greater than 0.01 and less than 0.05. One way of interpreting this value is that we would see a t-statistic of 2.09 or larger approximately 4 times in a 100 (or 1 time in 24), even if there was no difference between the means, by random chance alone. At this point you might be wondering firstly whether this is good enough to decide that the means are different, and secondly why this might be the case given that there was such a big overlap in the data. The answer to the first question is "it depends." It depends on what you decide is acceptable risk. Are you prepared to say that the means are different accepting that you will be wrong on average 5 times in 100? Or would you prefer a higher standard of proof - say fewer than 1 time in 100? A popular phrase among statisticians is "being a statistician means you never have to say you are sure about anything." This situation is the embodiment of that phrase. In many disciplines, the P-value is rigidly interpreted, in that any value that falls below $P = 0.05$ is "definitely significant," whereas anything that is above $P = 0.05$ is not. I prefer to put the significance in context. The confidence for the interval for the difference is from 5.7×10^{-7} to 2.9×10^{-5}. Acceptable variation from many studies of RIs for glass from a single source is about 4×10^{-5} [23]. Given the limits of the confidence interval is far smaller than this (by 1 to 3 orders of magnitude) I would probably regard these two means as not being sufficiently different to declare a difference. The answer to the second question is a little more difficult, and we will discuss it in the next section.

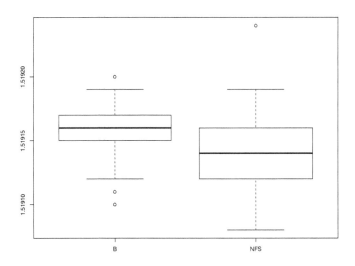

FIGURE 4.6: Box plot of RI measurements for the near float surface (NFS) and bulk (B) glass

4.5.6 The sampling distribution of the sample mean and other statistics

This section is important if you want to understand why the test statistics we use depend on both the variability in the data and the sample size. This section also will give a non-technical definition of the **Central Limit Theorem (CLT)**, which is used in many different situations to justify an assumption of normality. It is not essential that you read it now, but later when I wave my hands and say, "it does not matter because the CLT helps us here" you will know where to look.

Sampling from a distribution

Most people are familiar with the idea of sampling from a population. When we sample from a population, we randomly select a subset of individuals or objects. If we take a sample of data from a distribution, then the values are chosen in such a way that they occur with probability given by the pdf.

Now consider the following situation. We are going to conduct two experiments. The possible outcomes for each experiment are 1, 2, 3, or 4. Each outcome is equally likely. We are going to record the outcome for each experiment and the sum of both experiments. Firstly, let us look at the probability function for each experiment. There are four outcomes and they are equally likely; therefore, if X_1 and X_2 are random variables that record the outcome of each experiment, $\Pr(X_i = x) = 0.25, i = 1, 2$ and $x = 1, 2, 3, 4$. This probability function is represented graphically in Figure 4.7.

The question is, "What is the probability function for the sum?" That is, if $S = X_1 + X_2$, then what is $\Pr(S = s)$? Firstly, let us think about the possible outcomes. The smallest value that X_1 and X_2 can have is 1; therefore, the smallest sum is 2. Similarly, the largest value each experiment can have is 4, so the largest sum is 8. It should not be too hard to see that every other value between 2 and 8 is possible for the sum, so we wish to find $\Pr(S = s)$, $s = 2, 3, \ldots, 8$. The next question is, "Is every value of the sum S equally likely?" The answer is clearly no. For example, there is only one way to achieve a sum of 2, but there are three possible ways that we might get a sum of 4. Table 4.1 shows all the different outcomes that lead to the values that the sum might take on. We can now get the probability function for the sum by dividing the "# Ways" column by the total (16). This is shown in Figure 4.8. There are two things I want you to notice. Firstly, the sum, S, has its own probability function, which is different from the probability function for X_1 and X_2. Secondly, we have gone from a probability function that was flat to one that has an approximate "bell-shape." That is, the probability function of the sum is (very approximately) normal. The probability function will get more and more "normal" as we add together a larger and larger number of experiments. Figure 4.9 shows what happens to the distribution of the sum as we add 2 experiments, then 4 experiments, then 16 experiments,

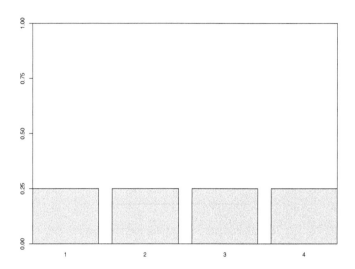

FIGURE 4.7: The probability function for the outcome of each experiment

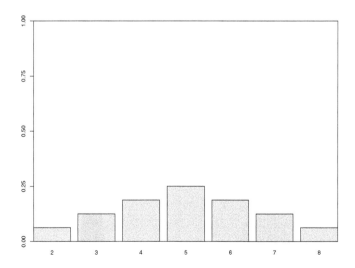

FIGURE 4.8: The probability function for S, the sum of the two experiments

S	X_1	X_2	# Ways
2	1	1	1
3	1	2	
	2	1	2
4	1	3	
	2	2	
	3	1	3
5	1	4	
	2	3	
	3	2	
	4	1	4
6	2	4	
	3	3	
	4	2	3
7	3	4	
	4	3	2
8	4	4	1
		Total	16

TABLE 4.1: Outcomes for the sum

and finally 64 experiments together. This is a demonstration of the Central Limit Theorem.

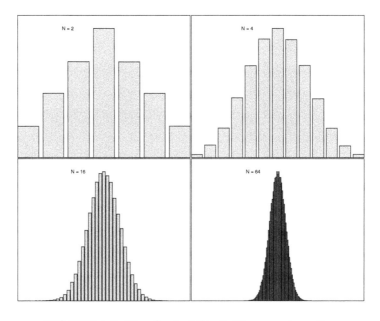

FIGURE 4.9: The Central Limit Theorem in action

Hypothesis tests and sampling theory

> **The Central Limit Theorem (CLT)**
>
> If $X_1, X_2, ..., X_n$ are independent random variables with the same distribution, then the sum of these variables $S = X_1 + X_2 + \cdots + X_n$ (and hence also the mean) will have an approximately normal distribution. This approximation will get closer as n increases.

Notice that the definition of the CLT does not say a single thing about the distribution of the random variables themselves. This is very important. It says, "the distribution of the (sample) mean is approximately normal no matter what kind of distribution our sample or our data come from."

> **Corollary to the CLT**
>
> If the standard deviation of the population (or distribution) is σ, then the standard deviation of the sample mean is $\sigma/\sqrt{n}$.
>
> We estimate the standard deviation of the sample mean by $s/\sqrt{x}$. This estimated standard deviation is known as the **standard error**.

> **Tip 16: What is so magical about $n = 30$?**
>
> Those of you who have done some statistics before may remember your instructor telling you that "if you have more than 30 data points then you can use the normal (distribution) tables." There is nothing inherently magical about $n = 30$. It is simply a reflection of the observation that if the data come from a symmetric unimodal distribution, or something near to that, then the CLT ensures that the distribution of the sample mean will be very close to normal for samples of size 30.

The standard error makes it easier to understand why the two sample t-test in our second example had a significant P-value, even though the range of the data overlapped. The t-test compares the difference in the sample means to the variability in the difference we might see if we could repeat our experiment many times. This variability is quantified by the standard error. We can see from Figure 4.10 that the intervals constructed with the standard errors do not overlap, hence the significant result.

102 *Introduction to data analysis with R for forensic scientists*

> **Tip 17: Hypothesis tests are a self-fulfilling prophecy**
>
> Have you heard anyone make this statement? What it means is that if we take a big enough sample, we will eventually reject **any** null hypothesis. You should be able to see that as the sample size n gets very large, the standard error will get very small (relative to the mean). Therefore, if we can take a big enough sample, we can reject any null hypothesis. That is, the tiniest of differences will eventually be significant. This is the reason why we should always quantify significant differences with a confidence interval.

4.5.7 The χ^2-test of independence

The χ^2 **(chi-squared) test of independence** is used to analyze tables of counts for two or more variables. That is, situations where we have two or more categorical variables for which we have counted the number of observations falling into the categories formed by the "crossing" of these variables. For example, Weinberg et al. [24] were interested in whether there were differences in various cranial features for Caucasians and African Americans. One of these features was the shape of the occipital squamous bone. Weinberg et

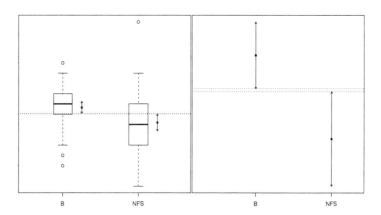

FIGURE 4.10: Examining the two sample t-test for bulk glass and near float surface glass in more detail

Hypothesis tests and sampling theory

al. classified the shapes into *narrow*, *equal*, or *greater*, depending on a relative width measurement. The cross-tabulated data are shown in Table 4.2. The variables have been **crossed** in that each of the cells in the table is represented by a category from each of the original variables.

	narrow	equal	greater
Afr. Amer.	2	14	9
Caucasian	17	9	6

TABLE 4.2: Occipital squamous bone classifications by race

I will assume in this section that we are talking about just two variables. However, the methods are easily extended to more than two variables. A χ^2-test may also be used to evaluate **goodness-of-fit** to an expected distribution under models other than independence. We will not cover this topic in this book. The questions that may be answered with a test of independence are:

- Are two variables associated or dependent?
- Is there a relationship between these two variables?
- Are the row frequencies or distributions the same for each group?
- Are the column frequencies or distributions the same for each group?

The χ^2-test can address all of these questions. It does so by looking at the cell frequencies and comparing them to the predicted or estimated frequencies. The estimated frequencies are calculated *under the assumption of independence*.

The independence rule

Two (random) variables, X and Y, are said to be **independent** if

$$\Pr(X \text{ and } Y) = \Pr(X) \times \Pr(Y)$$

Many people unconsciously (and often inappropriately) use the independence rule. Statistically, we write the null and alternative hypotheses as

$$H_0 : \pi_{ij} = \pi_i \pi_j \text{ for all values of } i \text{ and } j$$
$$H_1 : \pi_{ij} \neq \pi_i \pi_j \text{ for some values of } i \text{ and } j$$

where π_{ij} is the probability of a randomly selected observation "falling into" the ij^{th} cell in the table, or the jth column in the ith row.

Another way to understand the χ^2-test is to think about row frequencies. Imagine the categories on the rows describe different groups. In Table 4.2

this is easy to do as the different groups are the racial categorizations. If the row variable has no effect, then we would expect that *the frequencies in a particular column are (about) the same regardless of which row we are on.* In statistical terms, we would say that the *columns were independent of the rows.*

4.5.7.1 Example 4.3—Occipital squamous bone widths

Weinberg et al. [24] were interested in whether there were differences in shape of the occipital squamous bone for Caucasians. They recorded the possible shapes of the bone as *narrow, equal,* or *greater.* If there are significant differences in the different racial groups, then forensic anthropologists can use the occipital squamous bone as a discriminating variable for race. Inspection of Table 4.2 indicates there are sample differences, although this is more readily seen in Figure 4.11.

The R function `chisq.test` performs a χ^2-test of independence. The output from performing this test on Table 4.2 is given below.

```
        Pearson's Chi-squared test

data:   squamous.tbl
X-squared = 12.8634, df = 2, p-value = 0.001610
```

The *P*-value is of primary interest. A value of 0.0016 indicates that the differences between the observed values, and the values expected under the

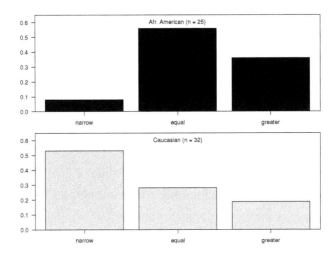

FIGURE 4.11: Occipital squamous bone category frequency by race

assumption of independence are too large to be explained by random chance alone. That is, there appears to be very strong evidence against the null hypothesis and that column probabilities do not change with the rows (and vice versa). The interpretation for a forensic anthropologist is that this measurement would be useful in classifying African American and Caucasian remains (as long as they included a skull).

The degrees of freedom in the χ^2-test relate to the number of rows and columns in the frequency table. It provides a useful check that the computer has treated the data correctly. If there are I row categories and J column categories, then the degrees of freedom should be $(I-1) \times (J-1)$. In our example, there are $I = 2$ row categories and $J = 3$ categories, so we can see that $(I-1) \times (J-1) = 2 \times 1 = 2$ is indeed the correct degrees of freedom figure.

The degrees of freedom actually relate to the number of cells in the table and the number of free parameters. We know that the row proportions must add up to one. Similarly, the column proportions must add to one. Therefore, if we *fix* one of the proportions so that this constraint will always be satisfied, then the other parameters may *vary freely*. It is not important to understand this. I have included this explanation for completeness.

4.5.7.2 Comparing two proportions

I commented earlier that it is possible to perform a hypothesis test and/or calculate a confidence interval for the difference between two proportions in a table as long as the populations and samples are independent. This means that we can compare proportions between populations but not within populations. To do this we need a test statistic and the distribution of the statistic under the null hypothesis. As long as the sample size is large enough, the distribution of a sample proportion, P, is approximately normal, with mean π, and standard deviation $\sqrt{\pi(1-\pi)/n}$. These values are estimated by the sample values $\hat{\pi} = p$ and $se(\hat{\pi}) = \sqrt{p(1-p)/n}$. Therefore, the difference in two independent proportions, P_1 and P_2, is also approximately normal with mean $\pi_1 - \pi_2$ and standard deviation

$$\sqrt{\frac{\pi_1(1-\pi_1)}{n_1} + \frac{\pi_2(1-\pi_2)}{n_2}}$$

These population values are estimated by replacing π_1 and π_2 with the sample values p_1 and p_2. Therefore, a $100 \times (1-\alpha)\%$ confidence interval for the difference in two proportions is given by

$$p_1 - p_2 \pm q(1-\alpha/2)\sqrt{\frac{p_1(1-p_1)}{n_1} + \frac{p_2(1-p_2)}{n_2}} \qquad (4.4)$$

where $q(1-\alpha/2)$ is the appropriate quantile from the standard normal distribution. Note that we do not use the t-distribution here as this is a large sample approximation. That is, it is approximately true as long as n_1 and n_2

are large enough. "Large enough" in this situation is when $n_i \times p_i > 5$ for both groups. If you wish to perform a hypothesis test rather than calculate a confidence interval, then the P-value can be calculated by calculating

$$Z_0 = \frac{p_1 - p_2}{\sqrt{\frac{p_1(1-p_1)}{n_1} + \frac{p_2(1-p_2)}{n_2}}}$$

and then finding $\Pr(Z \geq |Z_0|)$ using the cdf function for the standard normal. This can be done in R by typing `2*(1-pnorm(abs(Z0)))`, where Z0 is your test statistic Z_0.

4.5.7.3 Example 4.4—Comparing two proportions relating to occipital squamous bones

We can compare the proportion of African Americans and the proportion of Caucasians who have *equal*-sized occipital squamous bones. The sample proportions are $p_{AA} = 0.56$ and $p_C = 0.28$, respectively. The sample sizes are $n_{AA} = 25$ and $n_C = 32$. Therefore, a 95% confidence interval for the difference between the two proportions is given by

$$\begin{aligned} C.I. &= p_{AA} - p_C \pm q(0.975)\sqrt{\frac{p_{AA}(1-p_{AA})}{n_{AA}} + \frac{p_C(1-p_C)}{n_C}} \\ &= 0.56 - 0.28 \pm 1.96 \times \sqrt{\frac{0.56(1-0.56)}{25} + \frac{0.28(1-0.28)}{32}} \\ &= 0.28 \pm 1.96 \times 0.127 \\ &= (0.029, 0.528) \end{aligned}$$

Although it is not necessary to carry out a hypothesis test explicitly, because we can see that the confidence interval does not contain zero, we will do so for pedagogical reasons. The difference in the means is 0.28 and the standard error of the difference is 0.127, so our test statistic is

$$\begin{aligned} Z_0 &= \frac{p_{AA} - p_C}{se(p_{AA} - p_C)} \\ &= \frac{0.28}{0.127} \\ &= 2.19 \end{aligned}$$

Therefore, the P-value is $\Pr(Z > |Z_0|) = 2(1 - \Pr(Z > Z_0)) = 0.0284$. We reject the null hypothesis at the 0.05 level but not at the 0.01 level. A difference as large as this or larger would occur approximately 2.8% of the time or 1 time in 35 on average by random chance alone.

4.5.7.4 Example 4.5—SIDS and extramedullary haematopoiesis

Törő et al. [25] were interested in the frequency of extramedullary haematopoiesis (EMH) in the liver of sudden infant death (SIDS) cases. EMH has been linked to hypoxemia, which is decreased partial pressure of oxygen in the blood. They studied liver tissue samples from 51 SIDS cases and 102 non-SIDS controls. Immunohistochemistry showed eight cases in the SIDS group and six cases in controls with EMH. This data is shown in row frequency form in Table 4.3. In this study, the researchers were interested in whether EMH was related to SIDS, because an early diagnosis of EMH would lead to intervention and possible prevention of the child's death. Therefore, a hypothesis of interest would be that the probability of EMH is different in the SIDS cases versus the non-SIDS controls. The output from the χ^2-test is given below.

```
        Pearson's Chi-squared test

data:   liver.tbl
X-squared = 3.9311, df = 1, p-value = 0.0474
```

Note that the χ^2-test is performed on the table of counts, not the frequencies. The P-value in this case is quite close to 0.05, which indicates that we have evidence against the null hypothesis. In practical terms, this indicates that the evidence is not overwhelming. Furthermore, when you perform this test in R, you will find that you will get a warning from the procedure, and without some additional input, you will not get the output above. Why not?

4.5.7.5 Fisher's exact test

The χ^2-test is a **large sample approximation**. This means that in order for the test to give sensible results we need a relatively large sample size. You might notice that in our first example our total sample size was 57 ($= 25+32$) individuals. In our second example, we have a sample of 156 ($= 52 + 104$) livers. What is going on? The issue is really the expected cell counts rather than the total sample size. In the first instance, the χ^2-test will not work if any of the row or column sums are zero. Secondly, the test should not be used if one of the following is true:

- any cell has a zero count,

	EMH	No-EMH
Non-SIDS	0.06	0.94
SIDS	0.16	0.84

TABLE 4.3: Proportion of EMH diagnoses for non-SIDS ($n = 102$) and SIDS ($n = 51$) cases

- one or more of the expected cell counts is less than five in a table with two rows and two columns (a 2 × 2 table),

- more than 20% of the expected cell counts are less than five in a larger table.

R provides the expected counts when it performs the χ^2-test, and these are shown in Table 4.4. We can see from Table 4.4 that the expected cell count for the SIDS/EMH category is 4.67, and hence this is the reason that R has warned us. There are two common solutions. One is to perform **Yates's correction for continuity** [26] or **Fisher's exact test** [27]. R will automatically perform Yates's correction unless you tell it otherwise. The correction involves subtracting 0.5 from "the absolute value of the difference between the observed and expected values for each cell." Those of you who want to see the formula can look it up. Yates's correction dates from an era when computation was difficult. The same might be said of Fisher's exact test. However, Fisher's exact test has become more useful with the advent of modern computing.

Fisher's exact test can be used in situations where the conditions for the χ^2-test are not met. It is especially useful in situations where the total sample size is small or there are a large number of zeros in the table of counts. The test can be performed by the R function `fisher.test`. Fisher's test does tend to be conservative when the sample size is small. A conservative test is one that requires either a larger sample difference to detect an effect than a non-conservative test, or a larger sample size to detect the equivalent effect. Therefore, this conservativeness is desirable because it reflects the lack of certainty that comes with a small sample size.

Fisher's test was ignored for a long time because in its original implementation it required an exhaustive hand calculation that was not feasible for many problems—essentially any problem where there are more than two rows and two columns. This limitation has been overcome by using Monte Carlo solution. I will not go into the details; however, you need to be aware that Monte Carlo techniques do not return exactly the same answer every time. If they are good techniques, then the values from the test will be very similar.

4.5.7.6 Example 4.6—Using Fisher's exact test

The output from performing Fisher's exact test on the EMH data is given below.

	EMH	No-EMH
Non-SIDS	9.33	92.67
SIDS	4.67	46.33

TABLE 4.4: The expected cell counts for the EMH data under the assumption of independence

```
Fisher's Exact Test for Count Data

data:  liver.tbl
p-value = 0.07147
alternative hypothesis: true odds ratio is not equal to 1
95 percent confidence interval:
 0.09082937 1.18981497
sample estimates:
odds ratio
  0.338605
```

There are several things to notice, but the most interesting is that the P-value of 0.071 is not only larger than the P-value obtained from the χ^2-test (0.047) but also has moved the explanation from "evidence against the null hypothesis" to "weak evidence against the null hypothesis." There are a number ways of looking at this. In the first instance, the P-value from the χ^2-test was fairly close to the arbitrary 0.05 threshold. It would seem somewhat illogical to most forensic practitioners to do one thing if the P-value is just below the threshold and to do another if it is just over. In the evidence interpretation field, former U.K. forensic scientist Ken Smalldon called this the "fall-off-the-cliff" effect. It may be more sensible to try and quantify the effect and see whether it is "practically significant." The Fisher's exact procedure does this automatically for a 2×2 table with the **odds ratio**, or **OR**, as it is often abbreviated.

Recall that if the probability of an event A is given by $\Pr(A)$, then the odds on A can be obtained from

$$O(A) = \frac{\Pr(A)}{1 - \Pr(A)}$$

The odds ratio is, as the name suggests, a ratio of odds. The idea behind it is quite logical. If the odds on event A are $O(A)$ and the odds on event B are $O(B)$, then the odds ratio of A versus B is

$$OR = \frac{O(A)}{O(B)}$$

If the ratio is less than one, then this means that the odds of event B are lower than the odds of event A. If the ratio is greater than 1, then this means that odds of event A are higher than the odds of event B. If the odds are equal, then of course the ratio will be one. Note that we have talked about odds in each instance. If we talk in terms of probability, then we need to talk about **relative risk** or **likelihood ratios**. If the probability of event A in two mutually exclusives groups 1 and 2 are $\Pr(A|G_1)$ and $\Pr(A|G_2)$, respectively, then the relative risk of event A in group 1 versus group 2 is

$$RR = \frac{\Pr(A|G_1)}{\Pr(A|G_2)}$$

When we substitute the probabilities with **likelihoods**, which are defined in Chapter 6, the relative risk becomes a **likelihood ratio**. If the probability of the event is described by a discrete distribution, then the likelihood is equal to the probability and the relative risk is equal to the likelihood ratio.

The odds ratio for the EMH data gives us the odds of EMH occurrence in the non-SIDS group versus SIDS group. The sample estimate is 0.339. This means that odds of EMH in the non-SIDS group are about one-third of the odds of EMH in the SIDS group. Alternatively, we can take the reciprocal of the odds and say that the odds of EMH are about 3 times higher in the SIDS group compared to the non-SIDS group. The sample OR is an estimate; hence, it has some uncertainty associated with it. We can quantify this uncertainty with a confidence interval on the odds ratio. The confidence interval limits are from approximately 0.1 to 1.2. We read this as "the odds on EMH are from 10 times lower to 1.2 times higher." Given the vagueness of statement, it would probably be safer to err on the side of caution and say that the diagnosis of EMH is not particularly important in the prediction of SIDS deaths. In fact this was the conclusion of another paper (reference [4] cited in Töhö et al. [25]), which reported the same result—differences, but not statistically significant.

4.5.7.7 Example 4.7—Age and gender of victims of crime

Aitken et al. [28] were interested in whether there was a relationship between the gender and age of the victims in certain crimes. The age of the victims had been classified as 0–10 and 11+. The cross-tabulated data are shown in Table 4.5.

	0-10	11+	
Female	102	218	320
Male	46	43	89
	148	261	409

TABLE 4.5: Cross classification of victims of crime by gender and age

The output from performing a χ^2-test and Fisher's exact test is given below.

```
        Pearson's Chi-squared test

data:  sex.age.tbl
X-squared = 11.8344, df = 1, p-value = 0.0005815

        Fisher's Exact Test for Count Data

data:  sex.age.tbl
p-value = 0.0007466
alternative hypothesis: true odds ratio is not equal to 1
```

```
95 percent confidence interval:
 0.2637926 0.7262552
sample estimates:
odds ratio
 0.4383196
```

We can see that there are differences in the χ^2-test P-value (0.0006) and the Fisher's exact P-value (0.0007). However, none of these differences would lead us to change our conclusion of very strong evidence against the null hypothesis of no association. The confidence interval from Fisher's exact test tell us that the odds of a female victim being in the 0–10 age group were between approximately one-quarter (0.26) and approximately three-quarters (0.73) of the odds for a male victim.

4.6 Tutorial

1. Load the Bennett et al. data set.

    ```
    > data(bennett.df)
    ```

2. Recall this data consist of 49 groups of RI observations, with 10 observations per group. We will perform a two-sample t-test to see whether there is a difference between groups 1 and 3. In this data set, the different groups are in different columns of the data frame. The column names for bennett.df are X1, X2, ..., X49. Therefore, to perform a two sample t-test on groups 1 and 3 we can type

    ```
    > t.test(bennett.df$X1, bennett.df$X3)
    ```

    However, we should not do this without looking at our data first. There are only 10 observations per group, so a box plot is not appropriate. A simple scatter plot of the observations by group should suffice. To do this, we need to make a grouping variable using the rep command and we need to put the RI information for both groups into a single vector. We can perform these tasks by typing

    ```
    > group = rep(c(1, 3), c(10, 10))
    > ri = c(bennett.df$X1, bennett.df$X3)
    > plot(ri ~ group)
    ```

    You should print out group and ri to see what they look like. The plot is shown in Figure 4.12. We can see from the plot that the data lie in approximately the same position and they overlap considerably. Therefore, we can test the hypothesis that the mean RI of pane 1 is the same as the mean RI from pane 3. The R commands given below are equivalent and will give the same output. The second method is useful when your data is organized so that the measurements are in one column and the grouping information is in another column. It is a good idea to get used to organizing your data in this way as many of the procedures that we will encounter in later chapters need the data in this form.

```
> t.test(bennett.df$X1, bennett.df$X3)
> t.test(ri ~ group)
```

```
        Welch Two Sample t-test

data:  ri by group
t = 0.2507, df = 17.947, p-value = 0.8049
alternative hypothesis: true difference in means is not
equal to 0
95 percent confidence interval:
 -1.181040e-05   1.501040e-05
sample estimates:
mean in group 1 mean in group 3
       1.520010        1.520008
```

The *P*-value is very large here. We would conclude that there is no evidence of a difference between the true mean of pane 1 and the true mean of pane 3.

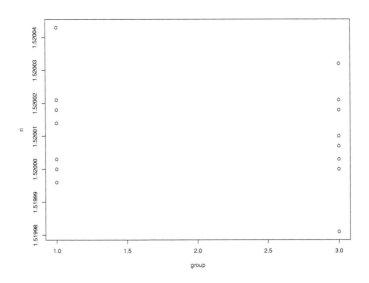

FIGURE 4.12: Plot of RI for the 1st and 3rd panels from the Bennett data set

3. Perform the above steps for groups 1 and 2.

    ```
    > ri = c(bennett.df$X1, bennett.df$X2)
    > group = rep(1:2, c(10, 10))
    > plot(ri ~ group)
    > t.test(ri ~ group)
    ```

4. Aitken et al. [28] were interested in whether there was a relationship between whether the victim had been abducted and the age of the victims in certain crimes. The age of the victims had been classified as 0–10 and 11+. This data is stored in the dafs library as abduct.age.df. Load the data and cross tabulate it by abduct and age.

    ```
    > data(abduct.age.df)
    > names(abduct.age.df)

    [1] "abduct" "age"

    > abduct.tbl = xtabs(~abduct + age, abduct.age.df)
    > abduct.tbl

           age
    abduct  0-10 11+
       No     92 198
       Yes    37  26
    ```

5. Perform a χ^2-test of independence to see whether there is an association between abduct and age.

    ```
    > chisq.test(abduct.tbl, correct = FALSE)

            Pearson's Chi-squared test

    data:  abduct.tbl
    X-squared = 16.2778, df = 1, p-value = 5.47e-05
    ```

6. We can examine the expected values used in the χ^2-test by storing the test result and extracting the expected values using $expected.

    ```
    > ct = chisq.test(abduct.tbl, correct = FALSE)
    > names(ct)

    [1] "statistic" "parameter" "p.value"   "method"
    [5] "data.name" "observed"  "expected"  "residuals"

    > ct$expected
    ```

```
             age
abduct         0-10       11+
   No    105.97734  184.02266
   Yes    23.02266   39.97734
```

7. Compare the results of the χ^2-test with Fisher's exact test.

```
> fisher.test(abduct.tbl)

        Fisher's Exact Test for Count Data

data:  abduct.tbl
p-value = 8.458e-05
alternative hypothesis: true odds ratio is not equal to 1
95 percent confidence interval:
 0.1788801 0.5925419
sample estimates:
odds ratio
 0.3276427
```

8. We can use a χ^2-test or Fisher's exact test to see whether there are differences in the genotype proportions of the different racial groups. We know that there are observable differences, but this exercise gives us the chance to confirm our observation with a test and also to see how to use the Monte Carlo version of Fisher's exact test.

```
> data(gc.df)
> gc.tbl = xtabs(~racecode + genotype, data = gc.df)
> gc.tbl
         genotype
racecode  AA  AB  AC  BB  BC  CC
       A   1  28   6 117  46   7
       C  19  24  60   4  29  63
       H   9  20  48  15  62  54

> chisq.test(gc.tbl)

        Pearson's Chi-squared test

data:  gc.tbl
X-squared = 287.6593, df = 10, p-value < 2.2e-16
```

The Monte Carlo version of Fisher's exact test requires us to tell R how many "samples" to take. In general, all Monte Carlo procedures become more accurate the bigger the sample size. Usually, we compromise between computational time and accuracy. For most Fisher's exact test, a sample size of 100,000 should be quick and sufficient:

```
> fisher.test(gc.tbl, simulate.p.value = TRUE,
        B = 100000)

    Fisher's Exact Test for Count Data with simulated
    p-value (based on 1e+05 replicates)

data:  gc.tbl
p-value = 1e-05
alternative hypothesis: two.sided
```

Note that Fisher's exact test does not return either an odds ratio or a confidence interval for the odds ratio in this example. This is because, although it is possible to construct an odds ratio, there is more than one odds ratio that is possible. For example, we might compare the odds of having the AA genotype in the African American population to the odds of having the AA genotype in the SW Hispanic population.

Chapter 5

The linear model

Remember that all models are wrong; the practical question is how wrong do they have to be to not be useful. – G. E. P. Box [29].

5.1 Who should read this?

The facetious answer to this question is "Everybody." However, this chapter is aimed at those readers how want to understand: how two or more variables may be related to each other; whether various factors in an experiment affect the response; or which variables may be important in predicting the behavior of another.

5.2 How to read this chapter

Some readers may be familiar with the concepts of regression, multiple regression, analysis of variance (ANOVA), and analysis of covariance (ANCOVA). These are all examples of linear modeling techniques. Most traditional statistics texts will breaks these topics into two or more chapters. However, the underlying mathematics, statistics, models, and methodology are absolutely the same. At a certain point, it starts to become labored to treat the topics differently. Having said this, there is no point in providing a completely non-standard treatment of the material. Therefore, in this chapter I will cover all of these topics. I will highlight the special language that is attached to the techniques, and I will try to indicate which are the most important things to concentrate on. There are also some very specialized linear models that allow us to examine complex designed experiments. The subject of experimental design will be in a separate chapter.

This chapter is also very long. For that reason I will include tutorial sections at the end of each major topic rather than at the end of the chapter. By doing this I hope that the reader will be able to focus on the material we have just covered rather than have to flip back many pages and refresh his or her memory.

5.3 Simple linear regression

Simple linear regression is the statistical description of a model we might use to describe the behavior of one continuous variable, Y, in relation to another continuous variable X. The model is

$$y_i = \beta_0 + \beta_1 x_i + \epsilon_i, \ \epsilon_i \sim N(0, \sigma^2), \ i = 1, ..., n \tag{5.1}$$

Recall that the equation for a straight line is $y = ax + b$, where a is the slope and b is the intercept. The simple linear regression model (5.1) says that each y_i is described by a straight-line relationship with x_i. The variables β_0 and β_1 are referred to as the **intercept** and **slope** in a simple linear regression model. However, in general, they are referred to as the **regression coefficients**. The variable ϵ_i is called the **error**. It is not an error or a mistake in the traditional sense. The statistical meaning of error is simply the difference between a particular observation and a mean. You may not recognize the mean in this model, but it is there. The difference is that the mean is described by a line rather than a single value. The final part of the model statement in equation 5.1 says that all the errors have the same distribution. That is, each error comes from a normal distribution with a mean of zero and a common variance of σ^2. This statement embodies the assumptions of independence, normality, and constant variance, which we need to make valid inferences. We will deal with these assumptions later.

The quantities, β_0, β_1, ϵ_i, and σ are all population quantities that we estimate from the data. For a particular data set, we have a **data model**

$$y_i = \widehat{y}_i + r_i \tag{5.2}$$

The values $\widehat{y}_i$ are called the **fitted values** and are calculated using the **fitted model**, which for simple linear regression is

$$\widehat{y}_i = \widehat{\beta}_0 + \widehat{\beta}_1 x_i$$

The estimated errors, r_i, are called the **residuals**.

There is some language that is useful to know for dealing with linear models. The variable Y is usually called the **response**, the **dependent** variable, or the **endogenous** variable. The variable X is called the **explanatory** variable, the **independent** variable, **predictor**, the **covariate**, or sometimes even the **exogenous** variable.

5.3.1 Example 5.1—Manganese and barium

We saw in Figure 3.14 that there appears to be a strong linear relationship between the concentration of manganese and the concentration of barium. We can explore this relationship with simple linear regression. The first step of any analysis like this should be to draw a plot; however, we have already seen this plot in Figure 3.14.

```
Call:
lm(formula = Mn ~ Ba, data = bottle.df)

Residuals:
   Min     1Q Median     3Q    Max
-5.512 -1.562 -0.225  1.248  5.341

Coefficients:
            Estimate Std. Error t value Pr(>|t|)
(Intercept)  12.0683     1.7865    6.76  5.7e-10 ***
Ba            0.2627     0.0101   26.00  < 2e-16 ***
---
Signif. codes:  0 '***' 0.001 '**' 0.01 '*' 0.05 '.' 0.1 ' ' 1

Residual standard error: 2.31 on 118 degrees of freedom
Multiple R-squared: 0.851,     Adjusted R-squared: 0.85
F-statistic:  676 on 1 and 118 DF,  p-value: <2e-16
```

We fit the simple linear regression model using the `lm` command in R. The output from this command is given above. The regression output contains a large amount of information. I will explain the various parts of the output, and then we will see how to use it in relation to our example.

1. The first two lines of output describe the `call`. This is a record for the analyst of the command that was issued to `lm`. We can use this to check that the output does indeed come from the model and data we wished to use. This can be of particular importance when we are using more than one model or more than one data set.

2. The next three lines give a **five number summary** of the residuals. A five number summary usually consists of the minimum, the maximum, the lower and upper quartiles, and the median. We can get an impression from this summary of the skewness of the residuals. Ideally, we would like the residuals to be symmetric around zero. We will come back to this point later on.

3. The next section is entitled `Coefficients`. We refer to this part of the output as the **regression table**. There are usually five columns to a regression table. These are

i. **Variable**: the name of the variable (or intercept).

ii. **Estimate**: the estimated coefficient. These estimates are usually written as $\widehat{\beta}_j$ and occasionally referred to as the *beta-hats* for obvious reasons.

iii. **Standard error of the estimates**: this column contains the estimated standard deviations (or standard errors) of the estimates. These values are written as $se\left(\widehat{\beta}_j\right)$. They are used to construct confidence intervals for the regression coefficients, and to test the hypotheses that a particular regression coefficient is zero. This is referred to as testing the **significance of a regression coefficient** or determining the **importance of variables**.

iv. **P-values**: this column contains the P-value for a test of the hypotheses

$$H_0 : \beta_j = 0$$
$$H_1 : \beta_j \neq 0$$

If the P-value is small, then this is evidence against the null hypothesis. That is, we believe that there is some evidence of a relationship between the particular explanatory variable and the response. R tags this final column with three asterisks * if $P \leq 0.001$, two asterisks if $0.001 \leq P \leq 0.01$, one asterisk if $0.01 \leq P \leq 0.05$, a period, or full-stop, if $0.05 \leq P \leq 0.10$, and nothing otherwise.

4. The residual standard error is an estimate of the standard deviation of the residuals. You may recall in equation 5.1 we wrote $\epsilon_i \sim N(0, \sigma^2)$. The residual standard error is an estimate of σ, i.e., $\widehat{\sigma}$.

5. The multiple R^2 and adjusted R^2 are measures of model performance. We will discuss their exact meaning shortly.

6. The final line of the regression output is an F-statistic and the associated P-value. This P-value is from a test of the **significance of the regression**. The hypotheses under consideration are

$$H_0 : \beta_1 = \beta_2 = \cdots = \beta_p = 0$$
$$H_1 : \text{some } \beta_j \neq 0, j = 1, \ldots, p$$

That is, we are testing the hypothesis that all of the regression coefficients except for the intercept (i.e., all of the slopes) are zero. In simple linear regression, we only have one slope, so this P-value is a test of the hypothesis $H_0 : \beta_1 = 0$. Therefore, this P-value will be exactly the same as the P-value on the slope in a simple linear regression. If we reject this hypothesis, then we say that the explanatory variable is **important** in predicting the response.

The first thing we look at in a regression is the *P*-value on the last line of the output. In this example, the *P*-value is very small. In fact R gives it as `< 2.2e-16`. This is the way that R represents "less than machine" precision. You can regard it as being "as close to zero as the computer can get" or just "extremely strong evidence against the null hypothesis." That is, we believe that there is very strong evidence for a linear relationship between the concentration of manganese and the concentration of barium in the bottles. The next statistic we look at is the value of R^2. This number is called the **squared multiple correlation coefficient**. It measures the (squared) correlation between the fitted values and the observed values. The **fitted values** or **predicted values** are the values we obtain by "plugging" the estimated regression coefficients into the model. For example, observation has a measurement of 50.2 for manganese and measurement of 149.9 for barium. The fitted predicted value of manganese for this observation using our estimate regression equation (or our fitted model) is

$$\widehat{Mn}_{86} = 12.0682759443172 + 0.26269911995605 \times 149.9$$
$$= 51.4468740257291 \approx 51.45$$

I have deliberately shown this in full precision so we do not have to worry (as much) about rounding errors.

5.3.2 Example 5.2—DPD and age estimation

Martin-de las Heras et al. [30] were interested in using deoxypyridinoline (DPD) crosslinks as a method of human age estimation. DPD is a non-reducible collagen crosslink that can be measured in human dentin samples extracted from permanent individual molars. Measurements of this protein were made in dentin samples from 22 patients with ages ranging from 15 to 73. A scatter plot of the data is shown in Figure 5.1. A smoothing line has been superimposed on Figure 5.1 and it would seem to indicate that there is slight curvature in the data. However, we will fit a straight-line model first and assess any improvements we might make later.

```
Call:
lm(formula = age ~ dpd.ratio, data = dpd.df)

Residuals:
    Min      1Q   Median      3Q      Max
-29.8056 -8.6384  -0.0186  8.4226  25.8918

Coefficients:
```

```
              Estimate Std. Error t value Pr(>|t|)
(Intercept)   10.83         7.78    1.39  0.17927
dpd.ratio     21.30         5.01    4.25  0.00039 ***
---
Signif. codes:  0 '***' 0.001 '**' 0.01 '*' 0.05 '.' 0.1 ' ' 1

Residual standard error: 14.3 on 20 degrees of freedom
Multiple R-squared: 0.475,     Adjusted R-squared: 0.448
F-statistic: 18.1 on 1 and 20 DF,  p-value: 0.000391
```

The output from fitting a simple linear regression model to the DPD data is given above. The regression is significant ($P = 0.0004$) meaning that there appears to be a significant relationship between `age` and `dpd.ratio`. The model is not fantastic at explaining the variation in age ($R^2 = 0.4746$). As we are in the simple linear regression situation, we are able to plot the fitted line on the data. This is shown in Figure 5.2. The straight line obviously cannot match the curvature but we can see that it is not too bad. These plots can sometimes be deceptive. Furthermore, we do not always have the luxury of plotting the fitted line on the data. Therefore, we need some tools that help us assess model fit. We base our assessment of model fit on a number of criteria, but the most important of these is meeting the assumptions of the linear model.

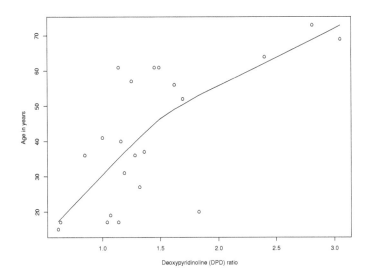

FIGURE 5.1: Scatter plot of age by deoxypyridinoline (DPD) ratio

The assumptions of the linear model

There is a core set of assumptions made about the linear model that are required to make statistical inferences. These assumptions are summarized in the model statement, by

$$\epsilon_i \text{ iid } N\left(0, \sigma^2\right)$$

where *iid* stands for *independently and identically distributed*. Explicitly the assumptions of the linear model are as follows:

1. The observations (and therefore the errors) are independent of one another

2. The errors have constant scatter (variance) around the line

3. The errors are normally distributed (with a mean of zero)

The assumptions of the linear model are listed in order of importance. Independence is the most important assumption and the hardest to test. There is no statistical remedy for lack of independence, although we can fit models to deal with dependency. These, however, fall into the realms of time series models, mixed effect models, or multivariate analyses, that are (mostly)

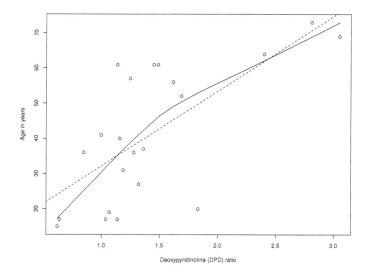

FIGURE 5.2: Age versus DPD ratio with fitted line (dashed)

beyond the scope of this book. We hope, in general, that independence is assured by the way in which the data were collected or the experiment carried out. Common causes for lack of independence are measurements on the same subject and time or space effects. For example, Gullberg [31] was interested in assessing the difference between six different breath alcohol analyzers. In his experiment, three volunteers provided $n = 10$ breath samples into each of six different instruments within an 18-minute time period. We expect there to be correlation or dependency in the measurements made on the same volunteer. Gullberg modeled this correctly using a variance components model.

The assumption of constant scatter is the next most important. If the errors have different variances, then the estimates of the regression coefficients will have unreasonably large standard errors for some values of y and standard errors that are too small for other values of y. The effect of this is that the significance tests that tell us whether a particular regression coefficient is zero or not will lack power. That is, it will lack the ability to detect a significant (non-zero) effect even when one is truly present. A common tool to detect non-constant scatter is to plot the residuals from the model against the fitted values. This plot is sometimes called a **pred-res plot**. We expect to see a broad homogeneous band of points centered around zero in the pred-res plot if the assumption of constant scatter is met.

Figure 5.3 shows three different situations. The fitted line plots are on the left-hand side and the associated pred-res plots are on the right-hand side.

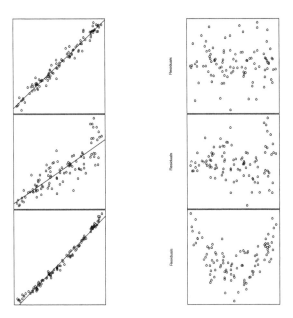

FIGURE 5.3: Residuaals versus predicted value (Pred-res) plots for three different regression situations

- In the first row, we see the ideal situation, namely, the errors are approximately equally distributed around the fitted line in the fitted line plot, and the zero line in the pred-res plot. There are no patterns in the pred-res plot.

- In the second row, we can see that the scatter increases around the fitted line as x increases. This effect is magnified in the pred-res plot. The residuals, however, are still symmetrically distributed around the zero-line in the pred-res plot. The data in this particular example were generated with $Var(\epsilon_i) \propto x_i \times \sigma^2$. That is, the variance increases in relation to the value of x

- In the third row, we can see that the data have slight curvature. This non-linearity is highlighted in the pred-res plot. When we see a pattern like this, we take it as evidence that we have not modeled all of the pattern or trend in the data. The data in this example where generated by using a quadratic model, i.e., $y_i = \beta_0 + \beta_1 x_i + \beta_2 x_i^2 + \epsilon_i$.

Pred-res plots are very good at magnifying non-constant scatter and non-linearity. This is shown to great effect in Figure 5.4. The data, which exhibits a quadratic trend are shown in the left-hand plot. The quadratic trend is very obvious in the pred-res plot on the right-hand side. Non-linearity and non-constant scatter can sometimes be dealt with by transforming the data. We will cover this topic later in this chapter. The pred-res plot for our simple

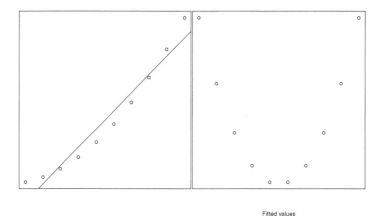

Fitted values

FIGURE 5.4: Pred-res plots magnify non-linearity. The quadratic trend here is magnified.

linear regression of age on dpd.ratio is shown in Figure 5.5. There are no noticeable patterns in this plot, although with such a small amount of data it would be hard to detect any. We do note that there is a large residual for the point with a fitted value of approximately 50. Closer investigation of this data reveals this to be a data point that has an estimated age of 49.7 years when the actual age of the subject was 20. It is worth investigating the effect of removing this data point from the analysis. In general, we try to avoid omitting data, but it is often worth taking a "try it and see approach." Of course, we must heed scientific rigor and make note of any data removed in our report or article. Removing this particular point increases R^2 to 0.5701.

```
            Estimate Std. Error t value   Pr(>|t|)
(Intercept)   10.037     7.0056  1.4327 0.16817850
dpd.ratio     22.868     4.5558  5.0196 0.00007609
```

The regression coefficients, on the other hand, have hardly changed. This means that this particular point simply has a large residual but not very much **influence** on the fitted model. No influence means that removing the point will have little to no effect on the predictions made from the model and therefore it is not worth removing.

The assumption of normality of the errors is the least important. There are numerous formal hypothesis tests in the literature for normality (e.g., Shapiro-Wilk, Anderson-Darling, Cramër-von Mises, etc.). However, although we statisticians make a big deal of normality in our writing and lecturing, we

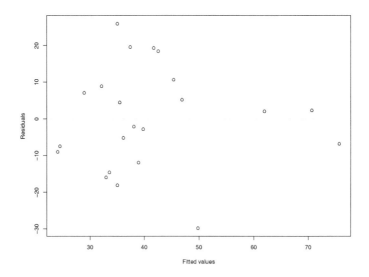

FIGURE 5.5: Pred-res plot for age versus dpd.ratio model

are often dismissive of it in practice. This is because in many situations we can rely on the Central Limit Theorem. As a rough rule of thumb, if we have more than 30 observations, then we can generally ignore departures from normality. Plotting methods are often a good way of assessing normality. We have two tools in our toolbox for carrying out this task: the normal Q-Q plot and a histogram (or density estimate) of the residuals. The latter is the simplest, since it involves computing the residuals, which is easily done using the `residuals` function in R and then applying the `hist` function to draw the histogram. It can also be useful to superimpose a density estimate. The former requires a little explanation.

5.3.2.1 The normal Q-Q plot

The **normal quantile-quantile plot** or **normal Q-Q plot** is a graphical tool for testing the normality of the data. It is primarily used as a diagnostic plot for linear models, but it is not restricted to this. The normal Q-Q plot is a scatter plot of the empirical quantiles from the data against the theoretical quantiles of the normal distribution. The points on the normal Q-Q plot will approximately follow a straight line if the data follow a normal distribution. The intercept of the line is an estimate of the mean, and the slope of the line is an estimate of the standard deviation of the associated normal distribution. Figure 5.6 shows common shapes for a normal Q-Q plot. Plot A shows truly

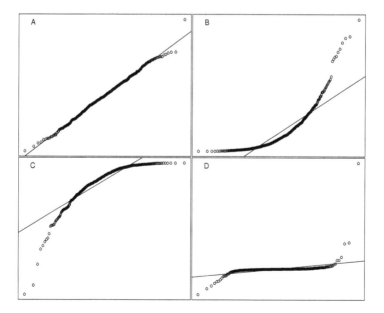

FIGURE 5.6: Possible shapes for a normal Q-Q plot. The theoretical quantiles are plotted on the x-axis and the empirical quantiles are plotted on the y-axis

normally distributed data. Note that there is a little *wobble* at each end of the normal Q-Q plot. This is a common feature of normal Q-Q plots even when the data are truly normally distributed. Plot B shows right-skewed data. Plot B shows left-skewed data, and plot D shows data where the tails are *heavier* or *fatter* than the normal distribution. The normal Q-Q plot is a large sample technique, meaning it is a relatively insensitive tool unless there is a large amount of data. However, given the robustness of the linear model to departures from normality this is not a cause for concern. It can be useful sometimes to compare the normal Q-Q plot for your data to five or six others created from samples of the same size drawn from a normal distribution. This can help you understand the variability you would expect to see in the plot. We will describe how to do this in the tutorial section of this chapter.

Returning to our example, we can now create a histogram of the residuals and a normal Q-Q plot for the residuals. Figure 5.7 shows the normal diagnostic plots for our DPD example. We have no reason to doubt the assumption of normality based on these plots.

5.3.3 Zero intercept models or regression through the origin

In many of the examples we encounter in this section, it is tempting to fit a model without an intercept. This is called a **zero-intercept model** or **regression through the origin** for the obvious reason that if $\beta_0 = 0$, then

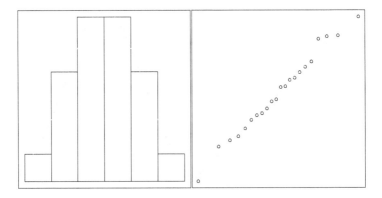

FIGURE 5.7: A histogram of the residuals and a normal Q-Q plot of the residuals from the DPD fitted model

the intercept is through the origin, and there is no reason to include it in the model statement. The temptation exists because there are many physical situations where we expect the response to be zero if the explanatory variable is zero. For example, we would expect the DPD ratio to be zero in newborn or unborn babies because they do not have teeth. However, we are not ever going to be making a prediction for this age group.

In general, the intercept does give us a value for the response when all of the explanatory variables are zero. It also reflects the variation due to variables not included in the model. We should not force the model through the origin (by omitting the intercept) unless we have a good theoretical reason for doing so *and* we have measurements near the origin. If we force the model through the origin when we do not have data near the origin, then we are effectively fitting a model beyond the range of the data. That is, we are making a prediction about the behavior of the response when we have no information about how it behaves.

In addition, R^2 has no meaningful interpretation in this context and therefore should not be used as a means of model evaluation. You should also note that the residuals do not necessarily sum to zero in this situation. The reader is refered to Eisenhauer [32] for a more thorough but readable discussion.

5.3.4 Tutorial

1. Load the bottle data and fit a simple linear regression model for manganese as a function of barium.

   ```
   > data(bottle.df)
   > mn.fit = lm(Mn ~ Ba, data = bottle.df)
   > summary(mn.fit)
   ```

2. Plot the fitted line on the data.

   ```
   > plot(Mn ~ Ba, data = bottle.df)
   > abline(mn.fit)
   ```

3. Add a smoothing line to the data as well, and color it blue.

   ```
   > with(bottle.df, lines(lowess(Ba, Mn), col = "blue"))
   ```

4. Plot the residuals versus the fitted values and a normal Q-Q plot of the residuals.

   ```
   > plot(mn.fit, which = 1:2)
   ```

5. Repeat the previous step, but this time extract the residuals and fitted values from the fitted object `mn.fit` using the `residuals`, `fitted` and `qqnorm` commands.

```
> res = residuals(mn.fit)
> pred = fitted(mn.fit)
> plot(pred,res, xlab = "Fitted values",
      ylab = "Residuals")
> abline(h = 0, lty = 2)
> qqnorm(res)
```

6. We use the `qqline` function in conjuction with `qqnorm` to construct a normal Q-Q plot with a straight line on it.

```
> res = residuals(mn.fit)
> qqnorm(res)
> qqline(res)
```

7. It is instructive to do the previous task by hand. To do this, we need to sort the residuals into order, and get the corresponding quantiles from the normal distribution. We can do this with the `ppoints` and `qnorm` commands.

```
> res = sort(residuals(mn.fit))
> n = length(res)
> z = qnorm(ppoints(n))
> plot(z, res, xlab = "Theoretical Quantiles",
      ylab = "Sample Quantiles", main = "Normal Q-Q Plot")
> abline(lm(res~z))
```

8. Plot a histogram of the residuals and superimpose the normal distribution on it.

```
> res = residuals(mn.fit)
> mx = mean(res)
> ## Note the residual standard error, not the simple
> ## standard deviation of the residuals is the estimate
> ## of the standard deviation of the residuals
> sx = summary(mn.fit)$sigma
> hist(res, prob = TRUE, xlab = "Residuals", main = "")
> x = seq(min(res)-0.5*sx, max(res)+0.5*sx, length = 200)
> y = dnorm(x, mx, sx)
> lines(x,y, lty = 2)
> box()
```

Note that we use the code `sx = summary(mn.fit)$sigma` instead of `sx = sd(res)` to estimate the standard deviation of the residuals, σ. This is because the residual standard error is the best estimate of σ.

9. The `s20x` package has a function that simplifies the last two steps called `normcheck`. `normcheck` can take either a single data set, a formula, or an `lm` object. So, for example, we can type

```
> normcheck(mn.fit)
```

The resulting graph is given in Figure 5.8. The normal Q-Q plot includes a Shapiro-Wilk test for normality, which you can ignore.

10. Plot a 95% confidence interval for the regression line for the mean and then for a new predicted value.

```
> plot(Mn ~ Ba, data = bottle.df)
> new.Ba = seq(min(bottle.df$Ba), max(bottle.df$Ba),
    length = 200)
> mn.pred.ci = predict(mn.fit, data.frame(Ba = new.Ba),
    interval = "confidence")
> mn.pred.pred = predict(mn.fit, data.frame(Ba = new.Ba),
    interval = "prediction")
> lines(new.Ba, mn.pred.ci[, 1])
> lines(new.Ba, mn.pred.ci[, 2], lty = 2, col = "blue")
> lines(new.Ba, mn.pred.ci[, 3], lty = 2, col = "blue")
> lines(new.Ba, mn.pred.pred[, 2], lty = 2, col = "red")
> lines(new.Ba, mn.pred.pred[, 3], lty = 2, col = "red")
```

You will notice that the prediction lines for a new value are much wider than for the mean. This is for the same reason that a confidence interval for a single value is much wider than a confidence interval for a mean.

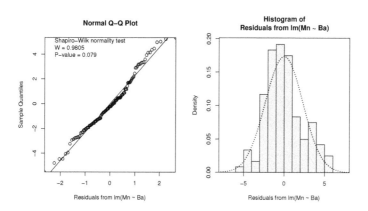

FIGURE 5.8: The `normcheck` function applied to the manganese and barium model

There is less variability in the average value. The question we should ask is, "Which one should we use?" To answer this question, we will predict an example value of Ba=180. If our aim is to answer the question "What is the average concentration of manganese in samples where the barium concentration is 180?" then we should give a prediction with a confidence interval for the mean, i.e.,

```
> new.Ba = 180
> mn.pred = predict(mn.fit, data.frame(Ba = new.Ba),
    interval = "confidence")
> mn.pred

       fit    lwr    upr
1  59.354 58.928  59.78
```

We would say "the average manganese concentration for samples with a barium concentration of 180 is 59.4 and it lies between 58.9 and 59.8 with 95% confidence." If the question is, "I have just collected a new sample and the barium value is 180. What is a reasonable predicted value and range for the manganese concentration?" then we should give a prediction interval for a new value, i.e.,

```
> new.Ba = 180
> mn.pred = predict(mn.fit, data.frame(Ba = new.Ba),
    interval = "prediction")
> mn.pred

       fit    lwr    upr
1  59.354 54.765 63.944
```

We would say "given the barium concentration is 180, we expect the manganese concentration in the new sample to be about 59.4 and it will lie between 54.8 and 63.9 with 95% confidence." Notice that both situations have the same predicted values. This is because, if $X_1, X_2, \ldots, X_n$ are an iid sample from some distribution then

$$E[X_i] = \mu$$

and

$$E[\overline{X}] = \mu$$

This says the expected value (mean) of a single value from a distribution and expected value (mean) of the sample mean of a sample from the same distribution are both the same.

11. Plot age versus the deoxypyridinoline (DPD) ratio in the DPD data set. Add a smoothing line to the plot. We will change the amount of smoothing because we have a relatively small amount of data.

```
> data(dpd.df)
> plot(age ~ dpd.ratio, data = dpd.df)
> lines(with(dpd.df, lowess(dpd.ratio, age, f = 0.95)))
```

12. Fit a regression model of age on dpd.ratio and use it to make a prediction for someone who as a dpd.ratio of 1.83.

```
> data(dpd.df)
> dpd.fit = lm(age ~ dpd.ratio, data = dpd.df)
> dpd.pred = predict(dpd.fit, data.frame(dpd.ratio = 1.83),
      interval = "prediction")
> dpd.pred
       fit    lwr    upr
1  49.806 18.929 80.682
```

Is this prediction useful? I think the answer is no. The range of the prediction interval (19–81) is almost wider than the range of the observed age data (11–73).

5.4 Multiple linear regression

Multiple linear regression models allow us to relate more than one explanatory variable to the response. That is, if we have p *linearly independent* explanatory variables, $X_1, X_2, \ldots, X_p$, which we think are related to a response variable Y, then we can attempt to model this relationship with

$$y_i = \beta_0 + \beta_1 x_{1i} + \beta_2 x_{2i} + \cdots + \beta_p x_{pi} + \epsilon_i, \quad \epsilon_i \sim N(0, \sigma^2) \quad (5.3)$$

We need a few extra tools to deal with models like this, but we have seen the vast majority of the technology we require already.

We have one guiding principle in this chapter, which is to "find the simplest model that best explains the data." This is often easier to state than it is to do.

5.4.1 Example 5.3—Range of fire estimation

Rowe and Hanson [33] conducted a study to test the validity of range-of-fire estimates to shotgun pellet patterns. A blind study was conducted in which questioned pellet patterns were fired at randomly selected ranges between 3.0 m and 15.2 m (10 and 50 ft) with two different 12-gauge shotguns, each firing a different type of buckshot cartridge. Test firings at known ranges were also conducted with the same weapons and ammunition. The data is shown in

Figure 5.9. We can see clearly from this scatter plot that the variation in the area increases in proportion to the range. There also appears to be slight curvature in the trend. A common statistical solution to deal with this pattern of behavior is to take the square root or logarithm of the response. The square root has a natural interpretation here in that the response is recorded in square inches (in^2). However, in this circumstance, it is not sufficient to remove the non-constant scatter. Taking logarithms, on the other hand, does seem to remove most of the problem. A plot of $log(area)$ versus range is shown in Figure 5.10. Note that the assessment is subjective. There are formal tests for homogeneity of variance, but they are susceptible to outliers. Formal tests of homogeneity are also a self-fulfilling prophecy in that as the sample size increases you will inevitability reject the hypothesis that the variances are equal, even if it is of no practical significance. We can now attempt to fit a linear model. Initially, we will explore a straight-line model.

```
Call:
lm(formula = log(area) ~ range, data = train.df)

Residuals:
    Min      1Q  Median      3Q     Max
-2.1704 -0.2047  0.0444  0.2429  0.8204

Coefficients:
            Estimate Std. Error t value Pr(>|t|)
```

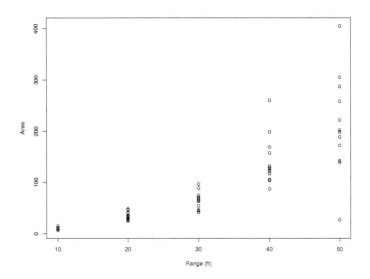

FIGURE 5.9: Pellet pattern area (in^2) versus firing range (ft)

```
(Intercept)    1.8651    0.1294    14.4    <2e-16  ***
range          0.0719    0.0039    18.4    <2e-16  ***
---
Signif. codes:  0 '***' 0.001 '**' 0.01 '*' 0.05 '.' 0.1 ' ' 1

Residual standard error: 0.428 on 58 degrees of freedom
Multiple R-squared: 0.854,     Adjusted R-squared: 0.852
F-statistic:  339 on 1 and 58 DF,  p-value: <2e-16
```

The regression output for the simple linear model is given above. We can see that there is very strong evidence that the linear trend is significant as the P-value on **range** is smaller than machine precision. The model also appears to be quite good with an R^2 of 0.8541. However, before we proceed we should examine the residuals and see whether any pattern remains. We can see in Figure 5.11 that there still appears to be some curvature in the data that we have not modeled. We can also see that there is a very large negative residual corresponding to one of the experiments conducted at 50 feet. We can address the curvature by trying a **quadratic model**. In mathematics a quadratic equation is one that takes the form $y = ax^2 + bx + c$. The behavior of the equation is shown in Figure 5.12. In most cases we do not expect data to behave completely like a quadratic equation. That is, we do not usually see parabolic shapes in data unless there is some particular physical property of the problem driving it. However, if we take some small segment of the quadratic, like the dashed line in Figure 5.12, then we can see how the

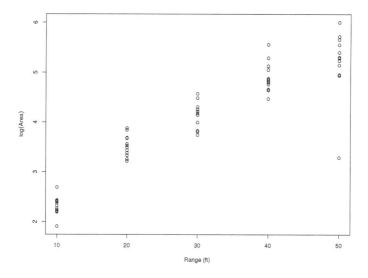

FIGURE 5.10: Log area versus firing range (ft)

136 *Introduction to data analysis with R for forensic scientists*

curvature in our data could be modeled over a small range by a quadratic term. Therefore, in our particular example, we do not expect the pellet scatter area

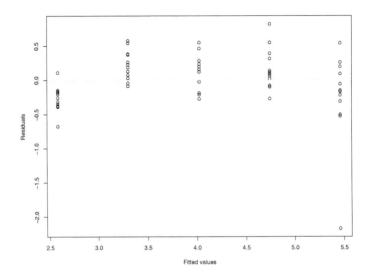

FIGURE 5.11: Pred-res plot for the shotgun scatter experiment

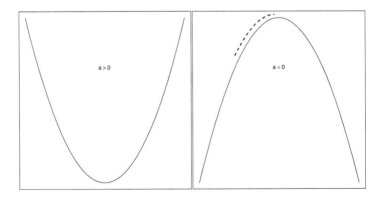

FIGURE 5.12: The shape of a quadratic

to start to decrease at some firing range, but we do think that it might stop increasing at a constant rate because of the effects of gravity. Also, we have to consider that at a certain range pellets simply will not reach the target. Therefore, we think we might be able to model the curvature we see in our data by adding a quadratic term.

```
Call:
lm(formula = log(area) ~ range + I(range^2), data = train.df)

Residuals:
     Min       1Q    Median       3Q      Max
-1.925518 -0.130353 0.000651 0.189302 0.789295

Coefficients:
             Estimate Std. Error t value Pr(>|t|)
(Intercept)  1.008104   0.233144    4.32 6.2e-05 ***
range        0.145353   0.017767    8.18 3.4e-11 ***
I(range^2)  -0.001224   0.000291   -4.21 9.0e-05 ***
---
Signif. codes:  0 '***' 0.001 '**' 0.01 '*' 0.05 '.' 0.1 ' ' 1

Residual standard error: 0.377 on 57 degrees of freedom
Multiple R-squared: 0.889,      Adjusted R-squared: 0.885
F-statistic:  228 on 2 and 57 DF,  p-value: <2e-16
```

The output from fitting a quadratic model is given above. There is one R "oddity" that we should address. When we wish to add a **polynomial term** of the form x^p to our model, e.g., $range^2$, then in R we must enclose it in the R interpretation inhibitor function I, e.g., I(range^2). It is important to understand that R will not return an error if you forget to do this, but in the same breath, neither will it do what you think it should. The model we have fitted here is called a **polynomial regression model** and is technically an instance of a **multiple linear regression** model, albeit a specialized one. It is important to note that $range$ and $range^2$ are linearly independent. They are obviously directly related to each other, but the relationship between them is quadratic, not linear; therefore, it is completely legitimate to add $range^2$ to a multiple regression model. We should check the model assumptions before we carry out any inference, such as determining the significance of the regression . The model assumptions are exactly the same as they were in the simple linear case. That is, we expect the errors to be independently and identically normally distributed with a mean of zero and a common variance of σ^2. Figure 5.13 shows that the behavior of the residuals is much less patterned than before. The normal Q-Q plot approximately follows a straight line, and the histogram of the residuals is approximately symmetric with the exception of one large outlier. Therefore, we can conclude that the model assumptions have now been met and proceed to interpret the regression output. The interpretation of the P-value associated with the F-statistic becomes more important in this setting. Recall that the null hypothesis is $\beta_1 = \beta_2 = \cdots = \beta_p = 0$. That is, none of the variables in the model is important in predicting the response. The P-value from the F-test for this hypothesis is smaller than machine precision. We interpret this as very strong evidence that "at least one of the regression coefficients (excluding the intercept) is non-zero" or alternatively that "the regression is significant." The next step is to assess which model terms are significant. We do this by looking at the P-values associated with each of the regression coefficients, excluding the intercept. Each of the P-values is extremely small. This indicates that there is very strong evidence that the regression coefficient for each terms is non-zero, or that the variables `range` and `range^2` are important in predicting the response. The appropriate measure of model fit in this circumstance is the adjusted R^2. The adjusted R^2 statistic is a way of making sure that models are not **over-parameterized**. One of the features of R^2 is that *will always increase with the addition of extra variables*. This means we can increase R^2 simply by adding variables to the model. One way of understanding this is to remember (or learn) that given

- two points we can draw a straight line through them

- three points we can draw a quadratic through them

- four points we can draw a cubic through them

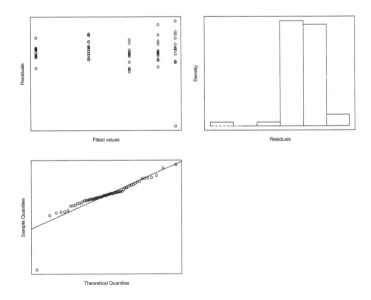

FIGURE 5.13: Diagnostic plots for the quadratic shotgun model

- and so on

This is demonstrated in Figure 5.14. The data in the six plots consist of seven *completely random* points. That is, there is no relationship between y and x. The fitted lines are a sequence of polynomial models. That is, the first model is $y = \beta_0 + \beta_1 x$. The second model is $y = \beta_0 + \beta_1 x + \beta_2 x^2$, and so on up to the last model, which is $y = \beta_0 + \beta_1 x + \beta_2 x^2 + \cdots + \beta_6 x^6$. At this point there is a term in the model for every single data point, and the data are described perfectly by the fitted line ($R^2 = 1$). However, given that we know there is absolutely no relationship between the variables, we know that this model is ridiculous. The adjusted R^2 is defined as

$$\text{adjusted } R^2 = 1 - (1 - R^2)\frac{n-1}{n-p-1} \quad (5.4)$$

This function penalizes the user for addition of variables to a model. It cannot be interpreted as "the percentage of variation explained," but it is useful when constructing a model. If the addition of a variable (substantially) increases the adjusted R^2, then we might consider that variable a useful addition to the model. Adjusted R^2 is always less than R^2, and it may be negative. Negative values of adjusted R^2 are an indication of severe over-parameterization of your model.

Returning to our example we can see that the adjusted R^2 has actually increased from 0.8515 to 0.8848. Is this a big increase? To put this in context,

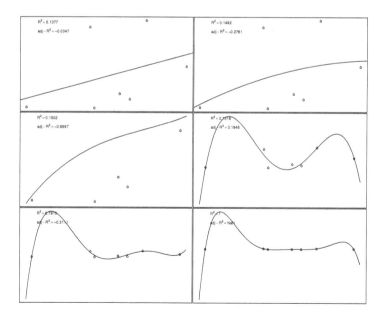

FIGURE 5.14: The dangers of over-parameterization. R^2 always increases with the addition of model terms.

we can contrast it against the removal of point 27. In this trial the shotgun was fired from 50 feet. The corresponding pellet area of $\approx 26.8\,\text{in}^2$ is comparatively low with respect to the mean of the other trials conducted at 50 feet ($\overline{Area}_{50} = 229.2\,\text{in}^2$, $\text{sd}(Area_{50}) = 79.48\,\text{in}^2$). If we omit this value model from our simple linear regression model the R^2 value increases to 0.9217. [1] There are no rules here, but I would argue that the improvement in fit does not outweigh the additional complexity that the quadratic model adds to the calibration task (which we have not covered yet).

5.4.2 Example 5.4—Elemental concentration in beer bottles

In this example we return to the bottle data. Recall that this data contain two variables relating to the object and location the measurements were made on (`Number` and `Part`), and the concentration of five elements (Mn, Ba, Sr, Zr, and Ti). We can see from the pairs plot (Figure 5.15) that there are strong linear relationships between pairs of the elements. It may be of interest to see whether we can relate the concentration of manganese to the other elements. In many ways this example is completely contrived. However, it will illustrate the mechanics and practice of multiple linear regression. It also reflects a common disparity between theory and practice. Most regression texts would have you believe that in every multiple regression data set there is one and

---

[1] The corresponding R^2 for the quadratic model without this point is 0.9434.

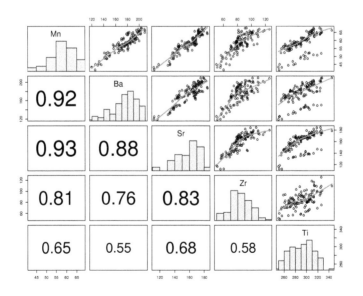

FIGURE 5.15: Pairs plot for bottle data

only one response variable, and the remainder of variables are explanatory. In modern data analysis this is often not the case. The choice for the response often depends on the question of interest, which may be motivated by some simple exploratory data analysis. Therefore, the model we will investigate is

$$Mn_i = \beta_0 + \beta_1 Ba_i + \beta_2 Sr_i + \beta_3 Zr_i + \beta_4 Ti_i + \epsilon_i, \quad \epsilon_i \sim N(0, \sigma^2)$$

```
Call:
lm(formula = Mn ~ Ba + Sr + Zr + Ti, data = bottle.df)

Residuals:
    Min      1Q  Median      3Q     Max
-5.2817 -1.2666  0.0514  1.2429  3.5914

Coefficients:
            Estimate Std. Error t value Pr(>|t|)
(Intercept)   1.9580     2.3782    0.82    0.412
Ba            0.1349     0.0162    8.32  2.0e-13 ***
Sr            0.1448     0.0272    5.33  5.0e-07 ***
Zr            0.0369     0.0194    1.91    0.059 .
Ti            0.0219     0.0106    2.07    0.041 *
---
Signif. codes:  0 '***' 0.001 '**' 0.01 '*' 0.05 '.' 0.1 ' ' 1

Residual standard error: 1.74 on 115 degrees of freedom
Multiple R-squared: 0.918,     Adjusted R-squared: 0.915
F-statistic:   322 on 4 and 115 DF,  p-value: <2e-16
```

The output from fitting this model is given above and the regression diagnostic plots are displayed in Figure 5.16. There is nothing in Figure 5.16 to give us any cause for concern. Looking at the regression output we can see that regression is definitely significant (P-value < 0.0001). The $R^2 = 0.9180$ and adjusted $R^2 = 0.9152$ are very high, indicating that the model does a good job of explaining the variation observed in the response. Inspection of the regression table shows that there is weak evidence ($0.05 < P < 0.10$) against the hypothesis that the coefficient on zirconium (Zr) is zero, and similarly there is only evidence ($0.01 < P < 0.05$) for titanium (Ti). These finding match the weaker correlations we observed in Figure 5.15 between manganese and these two variables. We can use the **anova** command in R to formally test the hypothesis $H_0 : \beta_3 = \beta_4 = 0$. There is evidence ($P = 0.0173$) against this hypothesis, as we might expect, but we should also compare the adjusted R^2 for the reduced model. As a quick aside the model with all of the variables in it is called the **saturated model**, and any model with some of the variables removed is called the **reduced model**. The reduced model (with the removal of Zr and Ti) has an adjusted R^2 of 0.9106. That is, there is essentially no change. The R^2 has gone from 0.9180 down to 0.9121, again essentially

Variable	Description	Range
Age	Real age in years	11–69
A	Attrition	0–3
P	Peridontosis	0–3
S	Secondary Dentine	0–3
C	Cementum Apposition	0–3
R	Root Resorption	0–3
T	Transparency	0–2

TABLE 5.1: Variables in Gustafson's teeth data [1]

no change. Therefore, following our guiding principle of finding the simplest model which best explains the data, we would choose the reduced model.

5.4.3 Example 5.5—Age estimation from teeth

Gustafson [1] was interested in predicting age at time of death from various dental features. He collected a data set of 41 teeth, which were scored on a number of variables. These are shown in Table 5.1. The score variables are integers between 0 and 3 (0–2 for transparency). There is a popular misconception that the explanatory variables must be continuous. This is completely false. The only restriction on the range of the variables is that the response should be approximately continuous. As always we proceed by

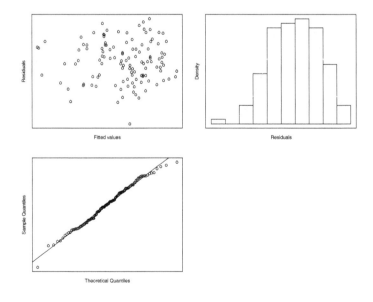

FIGURE 5.16: Diagnostic plots for the manganese model

exploring the nature of our variables with a pairs plot of the variables. We can see from Figure 5.17 that there is a reasonable correlation between Age and each of the explanatory variables. There are also some linear relationships between the explanatory variables, which may cause us problems. It can often be hard to see very much when the explanatory variables are heavily discretized, so we should be a little cautious in our interpretation of this plot. We will fit a linear model with all of the explanatory variables to this data, i.e.,

$$Age_i = \beta_0 + \beta_1 A_i + \beta_2 P_i + \beta_3 S_i + \beta_4 C_i + \beta_5 R_i + \beta_6 T_i + \epsilon_i,$$
$$\epsilon_i \sim N(0, \sigma^2)$$

```
Call:
lm(formula = Age ~ A + P + S + C + R + T, data = gustafson.df)

Residuals:
    Min      1Q  Median      3Q     Max
-10.393  -5.326  -0.567   3.125  12.917

Coefficients:
            Estimate Std. Error t value Pr(>|t|)
(Intercept)   12.57       2.72    4.62  5.3e-05 ***
A              3.49       1.74    2.01   0.0521 .
```

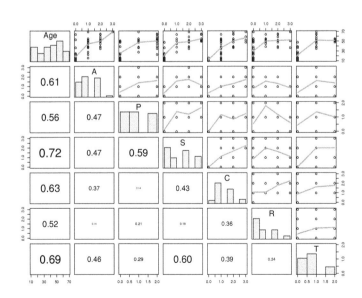

FIGURE 5.17: Pairs plot of the variables in Gustafson's teeth data

```
P              3.90         1.92    2.03    0.0504 .
S              2.99         1.57    1.91    0.0651 .
C              5.55         1.84    3.01    0.0049 **
R              4.38         1.27    3.46    0.0015 **
T              6.52         2.14    3.05    0.0045 **
---
Signif. codes:  0 '***' 0.001 '**' 0.01 '*' 0.05 '.' 0.1 ' ' 1

Residual standard error: 7.26 on 34 degrees of freedom
Multiple R-squared: 0.844,       Adjusted R-squared: 0.817
F-statistic: 30.7 on 6 and 34 DF,  p-value: 2.32e-12
```

We can see from the regression output that the regression is significant ($P < 0.0001$) and that the model explains a good proportion of the variation in the ages ($R^2 = 0.8443$). The adjusted $R^2 = 0.8168$ is close to the R^2, indicating the model is not over-parameterized. The regression diagnostic plots in Figure 5.18 show that there is no particular cause for concern with the model assumptions. We can see in the regression output that there are three variables, A, P, and S that are on the borderline of significance. There are many strategies and techniques for variable selection in regression modeling. We will employ a strategy called **backward elimination**. In this strategy we remove the least significant variable, i.e., the one with the largest P-value and re-fit the model. We take this approach rather than the wholesale removal of variables because a phenomenon known as **multicollinearity** may change the

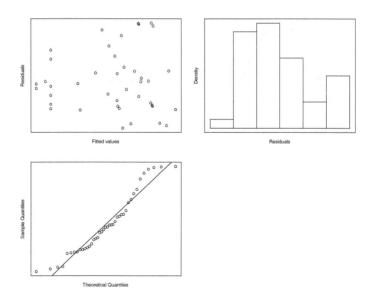

FIGURE 5.18: Regression diagnostic plots for teeth data

significance of other variables in the model. Multicollinearity occurs when two or more variables in the model are highly linearly correlated. The regression coefficients may change drastically with small changes in the model or the data when multicollinearity is present. Many authors and teachers of statistics make almost a fetish out of techniques for detecting multicollinearity. We will not do that but simply note that it is something we should be aware of.

The least significant variable is S.

```
             Estimate Std. Error t value   Pr(>|t|)
(Intercept)  11.8321      2.7917  4.2383 0.00015585
A             3.4530      1.7994  1.9190 0.06317156
P             5.8577      1.6869  3.4725 0.00139071
C             6.6744      1.8087  3.6902 0.00075714
R             4.0554      1.3031  3.1121 0.00368760
T             8.4023      1.9670  4.2716 0.00014130
```

We can see from the regression table above that the variable A has become the least significant where it was second least before and the variable P has become significant. The regression coefficient on the variable P has changed from 3.90 to 5.86. These are both indicators that multicollinearity is (or was) present. We will remove the variable A and re-fit the model again.

```
Call:
lm(formula = Age ~ P + C + R + T, data = gustafson.df)

Residuals:
   Min    1Q Median    3Q   Max
-11.42 -5.63  -1.51  3.61 19.50

Coefficients:
            Estimate Std. Error t value Pr(>|t|)
(Intercept)    12.39       2.88    4.31  0.00012 ***
P               7.21       1.59    4.54  6.0e-05 ***
C               7.60       1.81    4.20  0.00017 ***
R               3.77       1.34    2.81  0.00801 **
T               9.51       1.95    4.88  2.2e-05 ***
---
Signif. codes:  0 '***' 0.001 '**' 0.01 '*' 0.05 '.' 0.1 ' ' 1

Residual standard error: 7.81 on 36 degrees of freedom
Multiple R-squared: 0.809,     Adjusted R-squared: 0.788
F-statistic: 38.2 on 4 and 36 DF,  p-value: 1.70e-12
```

The regression diagnostic plots (not shown) exhibit slight right skew in the residuals. We could attempt to deal with this by taking logarithms of the response. However, the gain in the R^2 (0.8268 from 0.8095) is small compared to the additional model interpretation complexity that is introduced by doing so.

5.4.4 Example 5.6—Regression with derived variables

There are certain circumstances where it is advantageous to fit regression models with a set of **derived variables**. That is, instead of working with the original variables we use a set of new variables that are (linear) combinations of the original variables. For example, it is possible to perform a **principal components analysis** on the explanatory variables to find a smaller set of derived variables that represents most of the variation present in the original variables. We might do this to reduce the number of variables being fitted in the model. The Gustafson data contain an example of a derived variable. This variable, called TP, is the sum of the scores for the A, P, C, R, and T variables. Not surprisingly, TP is short for "total points." This summation across the other variables reduces the information to just a single score for each tooth, which may be sufficient to explain the variation in age.

	Estimate	Std. Error	t value	$\Pr(>|t|)$
(Intercept)	13.4506	2.2936	5.86	0.0000
TP	4.2612	0.3058	13.93	0.0000

TABLE 5.2: Simple linear regression of Age against TP (total points) using the Gustafson data

Table 5.2 shows the regression table for this model. The R^2 is 0.8327, which is a slight improvement over our simple linear regression model. This is probably due to the fact that the total points incorporate the information in the variables A and S.

Derived variables can sometimes be useful, but it is important to make sure that they can be explained sensibly and that they truly reflect the behavior of the data. We will test this in this example by picking a set of points across the range of Age, and compare the predicted values from the multiple regression model and the simple linear regression model.

The prediction intervals for both models are shown in Figure 5.19. The dotted horizontal lines represent the true ages. The prediction intervals for both models cover the true values. We can calculate the **mean squared prediction error** for each model as

$$MSE_{pred} = \frac{1}{n}\sum_{i=1}^{n}(y_i - \widehat{y}_i)^2$$

For the particular set of points used, the simple linear regression model using TP has a higher MSE_{pred} of 58.1 than the multiple linear regression model $MSE_{pred} = 40.1$.

5.4.5 Tutorial

1. Load the shotgun data and extract the training data set.

```
> data(shotgun.df)
> train.df =
    shotgun.df[shotgun.df$expt == "train",]
> names(train.df)

[1] "range"    "sqrt.area" "gun"      "expt"
```

2. Note that in this data set we have the square root of area but not area itself. We will use the `within` command to create a new variable `area` in the `train.df` data frame.

```
> train.df = within(train.df, {
    area = sqrt.area^2
  })
> names(train.df)

[1] "range"    "sqrt.area" "gun"      "expt"    "area"
```

3. Plot the area by range and the logarithm of area by range.

```
> par(mfrow = c(1, 2))
> plot(area ~ range, data = train.df)
> plot(log(area) ~ range, data = train.df)
```

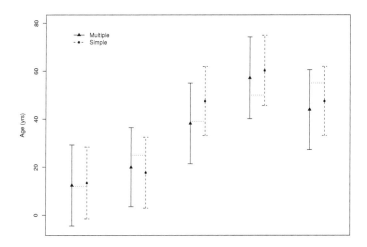

FIGURE 5.19: Prediction intervals for the Gustafson data using a multiple linear regression model and a simple linear regression model on a derived variable

4. Fit a simple linear model of `log(area)` on `range` and examine the residuals

   ```
   > train.fit = lm(log(area) ~ range, data = train.df)
   > plot(train.fit, which = 1:2)
   ```

   Note the curvature in the residuals. We might think that gravity is starting to have an effect on the shot as distances increases. Therefore, we can fit a quadratic term to the model, and see whether it improves the behavior of the residuals.

   ```
   > train.fit2 = lm(log(area)~range+I(range^2),
                 data = train.df)
   > plot(train.fit2, which = 1:2)
   ```

   The residuals are much better behaved now. It might be sensible to find the point with the large residual and take it out of the analysis. We can do this by applying the `which.max` function to the absolute value of the residuals. This will return the *index* or row number of the observation with the largest residual. We can omit the observation by putting a negative index in the `subs` option of the `lm` function.

   ```
   > bigRes = which.max(abs(residuals(train.fit2)))
   > bigRes
   ```

   27
   27

   ```
   > train.df[bigRes,]
   ```

   ```
      range sqrt.area    gun    expt    area
   27    50      5.18 Stevens  train  26.832
   ```

   ```
   > train.fit3 = lm(log(area)~range+I(range^2),
                 subs = -bigRes, data = train.df)
   ```

   That is, `subs = -bigRes`, tells R to leave observation number 27 out of the analysis.

5. Plot quadratic line on the data *on the original scale*. To do this we will have to back-transform to the original scale. Our model makes predictions for area on the logarithmic scale. Therefore, to back transform, we have to exponentiate the results with the `exp` function. This is because

$$\exp(\log(x)) = x$$

   That is, exponentiation *undoes* or *reverses* taking logarithms. We will also add 95% prediction interval lines.

The linear model

```
> new.range = seq(10,50,length = 200)
> train.pred = predict(train.fit3,
                       data.frame(range = new.range),
                       interval = "prediction")
> plot(area~range, data = train.df)
> lines(new.range, exp(train.pred[,1]))
> lines(new.range, exp(train.pred[,2]), lty = 2)
> lines(new.range, exp(train.pred[,3]), lty = 2)
```

This graph is shown in Figure 5.20.

6. Load the Gustafson data and plot the pairs plot. Before we do this we will drop the variables TN and TP as we do not want to use them in the analysis and they will complicate the later code.

```
> data(gustafson.df)
> gustafson.df = gustafson.df[, -c(1, 9)]
> pairs(gustafson.df)
```

You will note that this does not produce the nice plots I have used in the book. To get this we need to install and then load the s20x package. This is a set of functions that we give to our undergraduate class to simplify some of the R tasks.

This installation procedure is given for the dafs package on page 32.

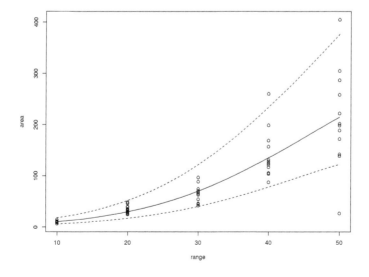

FIGURE 5.20: The fitted quadratic model on the original scale for the shotgun training data

The same instructions will work for the `s20x` package. Simply substitute `s20x` for `dafs` in the instructions. The package is updated two to three times a year, so you might repeat the installation procedure if you want to update. Once the package is installed, you need to load it to use it using the `library` function.

```
> library(s20x)
```

This needs to be done for each R session. That is, if you want to use the `s20x` library, each time you start R, you will need to use the `library` command. If you want to update the library, call the `install.packages` function *before* you call the `library` command.

Having done that we can use the `pairs20x` command from the `s20x` library.

```
> pairs20x(gustafson.df)
```

7. Fit the saturated multiple regression model for `age` versus all the other variables.

```
> g.fit = lm(Age ~ A + P + S + C + R + T, data = gustafson.df)
> summary(g.fit)
```

8. S is the least significant variable. We will use backward elimination to look at removing it. To do this, we will re-fit the model with the variable we wish to remove entering the model last. We compare this to a model with the variable removed, using the `anova` command. The F-test comparing the change in the sum of squares is called a **partial** F**-test**.

```
> g.fit = lm(Age ~ A + P + C + R + T + S, data = gustafson.df)
> g.fit2 = lm(Age ~ A + P + C + R + T, data = gustafson.df)
> anova(g.fit, g.fit2)

Analysis of Variance Table

Model 1: Age ~ A + P + C + R + T + S
Model 2: Age ~ A + P + C + R + T
  Res.Df  RSS Df Sum of Sq    F Pr(>F)
1     34 1793
2     35 1985 -1      -192 3.63  0.065 .
---
Signif. codes:  0 '***' 0.001 '**' 0.01 '*' 0.05 '.' 0.1 ' ' 1
```

This test tells us that the reduction in the sum of squares (of 191.64) is not significantly large enough to warrant the inclusion of S in the model. Therefore, the reduced model, without S is sufficient.

```
> signif(summary(g.fit2)$coef, 3)
```

```
            Estimate Std. Error t value Pr(>|t|)
(Intercept)   11.80       2.79    4.24 0.000156
A              3.45       1.80    1.92 0.063200
P              5.86       1.69    3.47 0.001390
C              6.67       1.81    3.69 0.000757
R              4.06       1.30    3.11 0.003690
T              8.40       1.97    4.27 0.000141
```

9. The regression table suggests that A is also not significant. We repeat the procedure above.

```
> g.fit2 = lm(Age ~ P + C + R + T + A, data = gustafson.df)
> g.fit3 = lm(Age ~ P + C + R + T, data = gustafson.df)
> anova(g.fit2, g.fit3)

Analysis of Variance Table

Model 1: Age ~ P + C + R + T + A
Model 2: Age ~ P + C + R + T
  Res.Df  RSS Df Sum of Sq    F Pr(>F)
1     35 1985
2     36 2193 -1     -209 3.68  0.063 .
---
Signif. codes:  0 '***' 0.001 '**' 0.01 '*' 0.05 '.' 0.1 ' ' 1
```

This confirms that we can remove A from the model.

```
> signif(summary(g.fit3)$coef, 3)
```

```
            Estimate Std. Error t value Pr(>|t|)
(Intercept)   12.40       2.88    4.31 1.22e-04
P              7.21       1.59    4.54 6.04e-05
C              7.60       1.81    4.20 1.65e-04
R              3.77       1.34    2.81 8.01e-03
T              9.51       1.95    4.88 2.17e-05
```

The revised regression table shows all of the remaining variables as being strongly significant.

5.5 Calibration in the simple linear regression case

Calibration and **inverse regression** are terms that are used to describe the process of constructing a **calibration line** or **curve**. In analytical chemistry, a calibration line is a way of determining the concentration of an unknown sample by comparing it to samples of known concentration. The calibration line itself is a regression line of the signal or response of some instrument on the concentrations of the known samples. The idea, however, easily extends beyond the realm of analytical chemistry. Both of the dental examples and the shotgun experiment can be thought of as calibration problems. A standard solution to this problem is to reverse the regression. That is, if the known standards have measurements x_i with associated response y_i, $i = 1, \ldots, n$, then if we wish to determine the measurement x for a new y, we can fit the model

$$x_i = \beta_0 + \beta_1 y_i + \epsilon_i, \quad \epsilon_i \sim N(0, \sigma^2)$$

and use this model to predict x. Interestingly, this is not the solution proposed by most of the chemometrics references that I have read. The classical solution estimates the x-value by regressing y_i upon x_i and then "solving for x." That is, if the estimated coefficients from fitting the model

$$y_i = \gamma + \delta x_i + \epsilon_i, \quad \epsilon_i \sim N(0, \sigma^2)$$

are $\widehat{\gamma}$ and $\widehat{\delta}$, then for predicted $\widehat{x}$ for a new value y is given by

$$\widehat{x} = \frac{y - \widehat{\gamma}}{\widehat{\delta}}$$

with a $100(1 - \alpha)\%$ confidence interval given by

$$\widehat{x} \pm \frac{\widehat{\sigma}_{y|x} t_{crit}}{\widehat{\delta}} \sqrt{\frac{1}{m} + \frac{1}{n} + \frac{(y - \bar{y})^2}{\widehat{\delta}^2 (n-1) Var(x)}}$$

where

- t_{crit} is the $100(1 - \alpha/2)\%$ critical point from a Student's t distribution on $n - 2$ degrees of freedom
- $\widehat{\sigma}_{y|x}$ is the residual standard error from the regression of y_i on x_i.
- m is the number of replicate measurements made on each standard.
- n is the total number of measurements made
- $\bar{y}$ is the mean of the y-values
- and $Var(x)$ is the sample variance of the x-values

This formula is given here for completeness. The way the formula works is best shown graphically. Figure 5.21 shows how the classical calibration formula produces a confidence interval for a value of x from a new value of y. The estimate itself is straightforward—we simply draw a horizontal line at the new y until it intersects with the regression line to get a predicted $\hat{x}$. The confidence limits for $\hat{x}$ are found by finding the intersection of the horizontal line on the upper and lower prediction limits of y_i given x_i. These prediction limits are not straight lines but slightly curved, hence the formulas given above.

Statisticians have been arguing for more than 40 years as to which method is preferable. However, a comprehensive book on the subject [34] and a recent paper that includes a literature review [35] indicate that although the problem is complex, the inverse regression method seems to be more reliable and produces "better" estimates. It is important to note that inverse regression methods should only be used when there is a very high degree of correlation between the response and explanatory variables. This is true of all of the following examples. However, in subject areas such as human age estimation this is usually not the case, and different alternatives must be explored (*Pers. Comm.—David Lucy*).

5.5.1 Example 5.7—Calibration of RI measurements

In forensic glass analysis, the Glass Refractive Index Measurer, or GRIM, instrument does not directly record refractive index. Instead, it records an average match temperature. Briefly, the glass fragments are mounted on a microscope slide in a droplet of silicone oil. The slide is placed into a Mettler hot-stage [36], and the temperature point at which the optical density of the oil

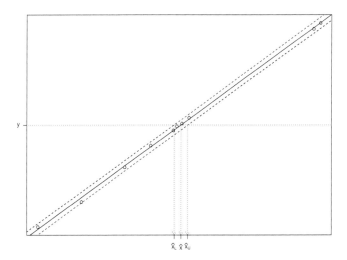

FIGURE 5.21: The classical calibration estimator and confidence interval

matches the glass fragment is determined by phase contrast microscopy. This match temperature is then converted into an RI by means of a calibration line. Bennett [37] performed a calibration experiment as a precursor to her spatial variation experiment. The manufacturers of the 2nd generation GRIM (called GRIM2), Foster and Freeman, recommend that a calibration experiment is performed each time a new bottle of Locke B Standard silicone oil is introduced to casework. The calibration of the instrument was established using standard glasses (Locke Scientific) set B1–B12 (RI=1.52912–1.520226). Each of the twelve reference sample glasses B1–B12 was measured five times. The data are shown in Figure 5.22. One feature that stands out in Figure 5.22 is the presence of an outlier around the 75 °C mark. This observation of 76.28 °C is in the set of measurements for the B7 standard. The other four measurements range from 79.43 °C to 79.61 °C. The group standard deviations of all the other standards range from 0.053 °C to 0.148 °C. The standard deviation for the B7 measurements is 1.453 °C. Omitting the outlier reduces this to 0.078 °C. Therefore, it would appear that something has gone wrong in this particular measurement, and it would be prudent to remove it from the subsequent analysis. The calibration line along with the 95% prediction interval is shown in Figure 5.23. The R^2 for this model is almost one, showing the incredibly good correlation between match temperature and refractive index. To test our method, we will predict the RI for the point that we omitted from our analysis. Recall that this observation had a match temperature of 76.28 °C and should have an RI close to that of the B7 standard, which is 1.51467. The inverse

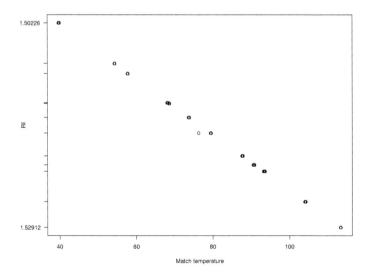

FIGURE 5.22: A calibration experiment for a GRIM2 instrument

regression model gives a prediction of 1.515801 and a 95% prediction interval of (1.515659, 1.515942). If we ignore the fact that each standard is measured five times (i.e., $m = 5$), then the classical method gives exactly the same estimate and prediction interval. However, if we incorporate this information, then the prediction interval is a little narrower (1.515735, 1.515866). Notice that neither of these intervals contains the RI of the B7 standard. On this basis we would reject the hypothesis that the erroneous measurement came from the B7 sample. Given that we know it did indeed come from the B7 standard we would be justified in thinking that something went wrong with the measurement of this fragment.

5.5.2 Example 5.8—Calibration in range of fire experiments

If you examine the shotgun data set (shotgun.df), you will see that there a variable called expt. This is a categorical variable that separates the measurements into values that are to be used for building the model (expt == train) and measurements that are to be used for testing the model (expt == predict). We built our model with the 60 training measurements. We can now test our model using the prediction measurements. Using the inverse regression method, and omitting one large outlier, our prediction model is

$$Range = -19.459 + 12.175 \log(Area)$$

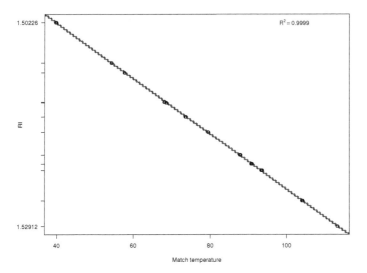

FIGURE 5.23: A calibration line for a GRIM2 instrument

This model has a high $R^2 = 0.9217$, and so should give good predictions. The predictions and associated prediction intervals are shown in Table 5.3

Actual	Predicted	Lower	Upper
14	10.6	2.44	18.82
19	23.1	15.02	31.17
22	24.1	16.02	32.16
30	34.7	26.59	42.72
36	37.2	29.13	45.29
37	36.5	28.44	44.59
38	44.5	36.32	52.59
44	42.3	34.14	50.37
45	49.0	40.84	57.23
45	45.1	36.95	53.24

TABLE 5.3: Predicted firing range (ft) from the shotgun experiment

and graphically in Figure 5.24. The 95% prediction intervals contain the true range value in every case. The average interval width is 16.2 ft with a standard deviation of 0.1 ft. This means that a scene examiner could predict the firing range to within ±8.1 ft with 95% confidence. The mean absolute difference between the predicted range and the actual range is 2.8 ft with a range of 0.1 ft to 6.5 ft.

5.5.3 Tutorial

1. We will compare the predictions made using the inverse method and classical method for the shotgun data. Firstly, load the shotgun data and extract the training data. We know from our previous work that observation 27 is problematic, and so we will remove it.

   ```
   > data(shotgun.df)
   > train.df =
       shotgun.df[shotgun.df$expt == "train",]
   > train.df = train.df[-27,]
   > train.df = within(train.df, {area = sqrt.area^2})
   ```

2. We will fit a simple linear regression model for log(area) on range, and the inverse model of range on log(area)

   ```
   > shotgun.fit = lm(log(area) ~ range, data = train.df)
   > invShotgun.fit = lm(range ~ log(area), data = train.df)
   ```

3. The data we wish to make predictions have the variable expt equal to predict in shotgun.df. We will extract them and store them in predict.df

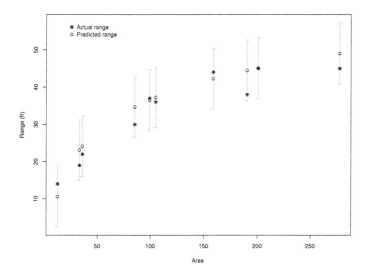

FIGURE 5.24: Predictions and prediction intervals for the shotgun experiment

```
> predict.df = shotgun.df[shotgun.df$expt == "predict", ]
> predict.df = within(predict.df, {area = sqrt.area^2})
> predict.df =
    predict.df[order(predict.df$range),]
```

4. The prediction intervals for the inverse regression are very simple to get using the `predict` command.

```
> invShotgun.pred = predict(invShotgun.fit,
                            data.frame(area = predict.df$area),
                            interval = "prediction")
> invShotgun.pred

         fit     lwr     upr
1     10.625  2.4350  18.815
2     23.092 15.0194  31.165
3     24.090 16.0213  32.158
4     34.659 26.5926  42.725
5     37.211 29.1331  45.289
6     36.512 28.4380  44.587
7     44.453 36.3151  52.590
8     42.257 34.1413  50.372
9     49.039 40.8440  57.234
10    45.097 36.9522  53.242
```

5. The classical method requires more work. We will set $m = 1$ even though we know $m = 5$. We will do this because it will make the intervals easier to compare with the inverse regression intervals.

```
> n = length(train.df$area)
> m = 1
> y = log(predict.df$area)
> my = mean(log(train.df$area))
> sigma = summary(shotgun.fit)$sigma
> vx = var(train.df$range)
> gamma = coef(shotgun.fit)[1]
> delta = coef(shotgun.fit)[2]
> t.crit = qt(0.975, n - 2)
> se = sigma/delta * sqrt(1/n + 1/m + (y - my)^2/(delta^2 *
       (n - 1) * vx))
> shotgun.pred = (y - gamma)/delta
> shotgun.pred = cbind(shotgun.pred, shotgun.pred -
       t.crit * se, shotgun.pred + t.crit * se)
> shotgun.pred

      shotgun.pred
[1,]       9.0088   0.46633 17.551
[2,]      22.5346  14.12447 30.945
[3,]      23.6167  15.21166 32.022
[4,]      35.0832  26.68066 43.486
[5,]      37.8523  29.43639 46.268
```

```
[6,]      37.0941 28.68237 45.506
[7,]      45.7088 37.22557 54.192
[8,]      43.3264 34.86819 51.785
[9,]      50.6844 42.13652 59.232
[10,]     46.4078 37.91649 54.899
```

6. The inverse regression method makes it easy for us to make predictions with a quadratic model.

```
> invShotgunQuad.fit = lm(range~log(area)+I(log(area)^2),
                data = train.df)
> invShotgunQuad.pred = predict(invShotgunQuad.fit,
              data.frame(area = predict.df$area),
              interval = "prediction")
```

7. It is probably easier to see the difference between the three sets of estimates graphically.

```
> xPos0 = 3*(1:10)-2
> xPos1 = xPos0 + 1
> xPos2 = xPos1 + 1
> ylim = range(range(shotgun.pred),range(invShotgun.pred),
            range(invShotgunQuad.pred))
> plot(xPos0, shotgun.pred[,1], pch = 20, col = "blue",
       ylim = ylim, xlim = c(0.5,20.5), axes = F,
       ylab = "Range (ft)", xlab = "")
> axis(2)
> arrows(xPos0, shotgun.pred[,2], xPos0,
         shotgun.pred[,3], angle = 90,
         length = 0.05, code = 3)
> points(xPos1, invShotgun.pred[,1], pch = 20,
         col = "red")
> arrows(xPos1, invShotgun.pred[,2], xPos1,
         invShotgun.pred[,3], angle = 90,
         length = 0.05, code = 3)
> points(xPos2, invShotgunQuad.pred[,1], pch = 20,
         col = "darkgreen")
> arrows(xPos2, invShotgunQuad.pred[,2], xPos2,
         invShotgunQuad.pred[,3], angle = 90,
         length = 0.05, code = 3)
> for(i in 1:10)
     lines(c(xPos0[i],xPos2[i]),rep(predict.df$range[i],2),
           lty = 2)
> box()
```

This code produces the graph shown in Figure 5.25 with colors.

5.6 Regression with factors

Sometimes, in addition to quantitative variables, we have categorical variables that we think might affect the response. For example, in the shotgun experiment, the authors [33] used two different shotguns. The reason for doing so might have been so the experimenters could discount the possibility that the results would be different for a different shotgun type. In fact Rowe and Hanson fitted a different model for each shotgun. The variable gun is called a **factor**. Each of the possible values that the factor may take on is called a **level** or a **factor level**. In the shotgun experiment, the factor gun has two levels: Remington and Stevens. The inclusion of one or more factor variables into a regression model has traditionally been called **analysis of covariance** or **ANCOVA**. However, this distinction is mostly unnecessary and old-fashioned, as the technology required to fit and evaluate these models is exactly the same as multiple regression. The only "trick" is knowing how to interpret the output.

Factor variables are added to regression models through the use of **dummy variables** or **variables**.

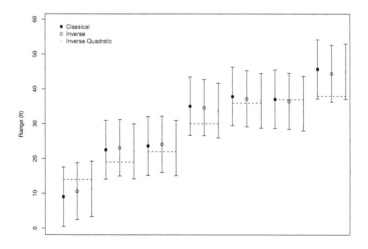

FIGURE 5.25: A comparison of the prediction intervals produced by the classical method and the inverse regression method for the shotgun data. The dashed line represents the true range.

Tip 18: Dummy variables for a k level factor

In general, $k - 1$ dummy variables are needed to represent a k level factor in a regression model.

The dummy variables are binary variables that take a value of zero or one, depending on the level of the factor. It is easy to see how this works for a factor with two levels. For example, the first 60 observations in the data set shotgun.df represent the training data we used to build out model. The first 30 measurements were made with a Stevens Model 77E 12-gauge shotgun, and the second 30 measurements were made using a Remington Model 870 12-gauge. The Stevens shotgun had a 20-inch barrel, whereas the Remington had a 12-inch barrel. This difference in barrel lengths might lead to a difference in the area as we might expect a wider spread from a shorter barrel. We can include d as a dummy variable in our model. It receives a value of zero ($d = 0$) if the Remington model was used and one ($d = 1$) if the Stevens shotgun was used. The simplest model we can fit with a dummy variable is called a **different intercepts** model. This model fits a different intercept for each level of the factor. For example, in the shotgun experiment, we can fit the model

$$Range_i = \beta_0 + \beta_1 d_i + \beta_2 \log(Area_i)$$

If the Remington shotgun was used, then $d_i = 0$ and so the model becomes

$$Range_i = \beta_0 + \beta_1 \times 0 + \beta_2 \log(Area_i)$$
$$= \beta_0 + \beta_2 \log(Area_i)$$

If the Stevens shotgun was used, then $d_i = 1$ and the model becomes

$$Range_i = \beta_0 + \beta_1 \times 1 + \beta_2 \log(Area_i)$$
$$= (\beta_0 + \beta_1) + \beta_2 \log(Area_i)$$
$$= \beta_1^* + \beta_2 \log(Area_i)$$

β_0 and β_1 are both constants. That is, they do not change with the value of $Area$. Therefore, they can be combined into a single new constant β_1^*, which is the intercept for the Stevens experiments.

The situation becomes a little trickier when there are more than two levels. Hypothetically, imagine that the shotgun experiments also had some trials using a Browning A-5 12-gauge shotgun. It is impossible to represent all three levels of the **gun** factor with one binary dummy variable. To do this, we need two dummy variables. This works in the following way: $d_1 = d_2 = 0$

(gun = SRemington), $d_1 = 1$, $d_2 = 0$ (gun = Stevens) and $d_1 = 0$, $d_2 = 1$ (gun = Browning). The regression model is

$$Range_i = \beta_0 + \beta_1 d_{1i} + \beta_2 d_{2i} + \beta_3 \log(Area_i)$$

It should be easy to see that the intercepts are

gun	d_1	d_2	Intercept
Remington	0	0	β_0
Stevens	1	0	$\beta_0 + \beta_1$
Browning	0	1	$\beta_0 + \beta_2$

One question you might ask is, "Why not just fit a different model for each group of data?" It is certainly possible to do this. However, we gain some "strength" by analyzing the data simultaneously, in that there is more data available to estimate the residual standard error. This means that our prediction intervals will be narrower if we combine the data rather than fit separate models.

We do not have to restrict ourselves to keeping the slope constant in this framework. That is, we can fit a **different slopes** model. Note that most people do not constrain the intercept and allow the slopes to differ, so a different slopes model usually implies that the intercepts are allowed to vary as well. To fit such a model, we include an **interaction** term. In general, if there is an interaction between a factor variable and a quantitative variable, then this means that the quantitative variable behaves differently for different levels of the factor. The different slopes model for our shotgun experiment is

$$Range_i = \beta_0 + \beta_1 d_i + \beta_2 \log(Area_i) + \beta_3 d_i \times \log(Area_i)$$

The term $d_i \times \log(Area_i)$ represents the interaction between the gun and the area. If **gun=Remington**, then $d = 0$, and so the model becomes

$$Range_i = \beta_0 + \beta_1 \times 0 + \beta_2 \log(Area_i) + \beta_3 \times 0 \times \log(Area_i)$$
$$= \beta_0 + \beta_2 \log(Area_i)$$

Similarly, if **gun=Stevens**, then $d = 1$, and so the model becomes

$$Range_i = \beta_0 + \beta_1 \times 1 + \beta_2 \log(Area_i) + \beta_3 \times 1 \times \log(Area_i)$$
$$= (\beta_0 + \beta_1) + (\beta_2 + \beta_3) \log(Area_i)$$
$$= \beta_1^* + \beta_2^* \log(Area_i)$$

We can see that the model has a different intercept and slope for each type of shotgun as desired.

5.6.1 Example 5.9—Dummy variables in regression

We will fit a different intercepts model to the shotgun data. It is important to note that we do not have to construct the dummy variables ourselves. R will automatically generate them based on the levels on the factor variable. We can see from the regression table (Table 5.4) that there is weak evidence for a difference between the two different shotguns. The fitted model is shown in

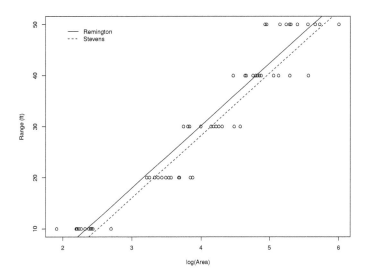

FIGURE 5.26: A different intercepts model for the shotgun experiment

Figure 5.26. In evaluating the utility of this model, we must ask two questions that are related. Firstly, "Does the new model explain a significantly more of the variation in the response?" and secondly, "Does the new model improve our predictions?" The first question can be answered by comparing the R^2 values of the models. These are 0.9217 and 0.9265 respectively. A half percent gain hardly seems worth it. We can only answer the second question if we are prepared to assume that the shotgun type is known. This seems a reasonable assumption if the only fact under question is the distance from which the

	Estimate	Std. Error	t value	$\Pr(>|t|)$
(Intercept)	-18.6689	1.9657	-9.50	0.0000
log(area)	12.2159	0.4599	26.56	0.0000
gunStevens	-1.9416	1.0168	-1.91	0.0613

TABLE 5.4: Regression table for the shotgun data with a different intercepts model

shotgun was fired. The average confidence interval width is 16 ft with a standard deviation of 0.09 ft. The mean absolute difference is 2.4 ft, with a range of 0.5 ft to 5.5 ft. There are very slight gains in prediction accuracy, but none is large enough to warrant using the extra information.

5.6.2 Example 5.10—Dummy variables in regression II

As we noted in our GRIM2 calibration example, each new bottle of silicone oil necessitates a new calibration experiment. Therefore, laboratories that do glass work will often have calibration lines that differ from year to year. Dr. Grzegorz Zadora from the Institute for Forensic Research in Krakow, Poland, has generously provided data from a GRIM2 calibration experiment. We can combine this data with the calibration data from Bennett [37] to investigate the hypothesis that these two experiments will yield different calibration lines. That is, we can use the factor owner in the ri.calibration.df data set to fit a different slopes model. The factor owner has two levels, RB for Rachel Bennett, and GZ for Grzegorz Zadora.

```
Call:
lm(formula = ri ~ temp * owner, data = ri.df)

Residuals:
     Min         1Q     Median         3Q        Max
-6.18e-04  -5.54e-05   9.64e-06   6.63e-05   1.98e-04

Coefficients:
                Estimate Std. Error  t value Pr(>|t|)
(Intercept)     1.54e+00   5.23e-05 29483.08  < 2e-16 ***
temp           -3.62e-04   6.55e-07  -552.29  < 2e-16 ***
ownerRB         3.62e-04   7.45e-05     4.86 3.7e-06 ***
temp:ownerRB   -2.69e-06   9.30e-07    -2.89  0.0046 **
---
Signif. codes:  0 '***' 0.001 '**' 0.01 '*' 0.05 '.' 0.1 ' ' 1

Residual standard error: 0.000105 on 115 degrees of freedom
Multiple R-squared:     1,      Adjusted R-squared:     1
F-statistic: 2.03e+05 on 3 and 115 DF,  p-value: <2e-16
```

Before we interpret the output, it is worth briefly looking at the Call. The call is the command that I gave to R fit the model. In the call you can see formula = ri ~ temp * owner. This formula seems to have only one term in it temp * owner when we want to fit the model

$$RI_i = \beta_0 + \beta_1 d_i + \beta_2 Temp_i + \beta_3 d_i \times Temp_i$$

which has three terms in it in addition to the intercept. As you might suspect, the model is written in a compact notation. This is called the Wilkinson-

Rodgers notation [38], and it provides us with a way of efficiently specifying potentially complex models. If we look further down the output to the regression table, we can see all of the terms we expect to see. The intercept offset β_1 is labeled `ownerRB`. This means that if we wish to find the intercept for Bennett's data we add the estimated coefficients for (`Intercept`) and `ownerRB` together. Similarly, if we wish to find the slope for Bennett's data, we add the coefficients for `temp` and the interaction `temp:ownerRB` together. By default, R deals with factor levels alphabetically. So, although we have been talking about Bennett's data followed by Zadora's, GZ, of course, comes before RB. If it is important to deal with factor levels in a particular order, then it is possible to reorder the levels through the `as.factor` command. We will deal with this specifically when we discuss analysis of variance.

Returning to the regression output, we can see that the regression is definitely significant ($P \approx 0$), and that all the regression coefficients have P-values less than 0.01, which indicates there is strong evidence for the different slopes model. Figure 5.27 shows that the residuals are slightly left skewed. This is not really an issue as given that we have 119 observations we can rely on the Central Limit Theorem. The largest residual is from observation 108, which has a temperature measurement of 37.31 °C. Comparing this measurement to the mean and standard deviation of the other measurements in this group shows it to be 5.6 standard deviations below the mean. Removal of this data point improves the R^2 by a minor amount and the residual standard error. Interestingly, it reduces the evidence for different slopes in the model

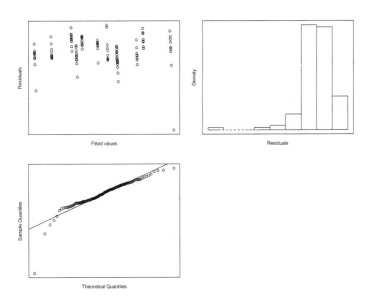

FIGURE 5.27: Regression diagnostic plots for the RI calibration model

$(0.01 < P < 0.05)$. This is perhaps not surprising as outliers that are a long way from the center of the data (i.e., a long way from $(\overline{x}, \overline{y})$) will influence the fit. A point like this is said to have high **leverage**. Figure 5.28 shows the fitted lines (without point 108). Although the lines are very close together we can see the deviation due to the different slopes as the match temperature decreases the evidence for the different slopes model.

5.6.3 A pitfall for the unwary

It is very important to make sure that there are enough observations for each level of the factor. There are two reasons for this. In the first instance, when you include a factor in your model, you are fitting a different mean to the different groups specified by the levels of the factor. In order to get a sensible estimate of that mean, you will need a reasonable amount of data. As a rough rule of thumb, approximately 10–15 observations would be a minimum. Any fewer than that and you should take great care in your interpretation and any generalizations you might make. The second reason is that it is very easy to over-parameterize when you include a factor in your model. The reason is that the $k-1$ dummy variables for the factor levels use up an addition $k-1$ degrees of freedom (or $k-1$ observations). If k is large relative to the number of observations n, then you can very quickly find yourself running out of data, especially if your model contains interactions.

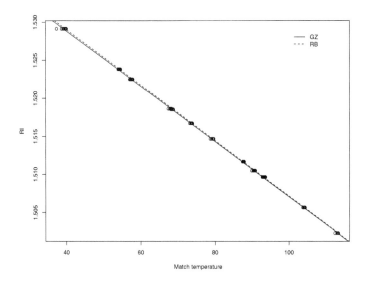

FIGURE 5.28: Different slopes model for GRIM2 calibration

5.6.4 Tutorial

1. Load the GRIM2 calibration data set, remove points 31 and 108, and rename the data set `ri.df`.

   ```
   > data(ri.calibration.df)
   > ri.df = ri.calibration.df[-c(31, 108), ]
   ```

2. Fit the different slopes model to this data with RI as the response.

   ```
   > ri.fit = lm(ri ~ temp * owner, data = ri.df)
   ```

3. Check the model assumptions for this data.

   ```
   > plot(ri.fit, which = 1:2)
   > res = residuals(ri.fit)
   > hist(res, prob = TRUE)
   > mx = mean(res)
   > sx = summary(ri.fit)$sigma
   > x = seq(min(res) - 0.5 * sx, max(res) + 0.5 * sx,
       length = 200)
   > lines(x, dnorm(x, mx, sx), lty = 2)
   ```

4. Write down the fitted model for Grzegorz's data and the model for Rachel's data. To do this we need the coefficients from the regression table. We can use the `summary` command, or we can use the `coef` command to extract the coefficients directly.

   ```
   > b = coef(ri.fit)
   > b
   ```

   ```
     (Intercept)           temp        ownerRB   temp:ownerRB
      1.5432e+00    -3.6123e-04     3.9158e-04     -3.1425e-06
   ```

   Remember that R vectors start at index 1, so $\widehat{\beta}_0$ will be stored in `b[1]`, $\widehat{\beta}_1$ in `b[2]` and so on. The model for Grzegorz's data is given by setting the dummy variable `ownerRB` to zero, i.e.,

   $$\widehat{ri}_i = \widehat{\beta}_0 + \widehat{\beta}_1 \times 0 + \widehat{\beta}_2 temp_i + \widehat{\beta}_3 \times 0 \times temp_i$$
   $$= \widehat{\beta}_0 + \widehat{\beta}_2 temp_i$$

   where $\widehat{\beta}_0$ =`b[1]` and $\widehat{\beta}_2$ =`b[3]`. We get Rachel's model by setting ownerRB to one, i.e.,

$$\widehat{ri}_i = \widehat{\beta}_0 + \widehat{\beta}_1 \times 1 + \widehat{\beta}_2 temp_i + \widehat{\beta}_3 \times 1 \times temp_i$$
$$= \widehat{\beta}_0 + \widehat{\beta}_1 + \widehat{\beta}_2 temp_i + \widehat{\beta}_3 temp_i$$
$$= (\widehat{\beta}_0 + \widehat{\beta}_1) + (\widehat{\beta}_2 + \widehat{\beta}_3) temp_i$$
$$= \widehat{\beta}_0^* + \widehat{\beta}_1^* temp_i$$

where $\widehat{\beta}_0^*$ =b[1]+b[2] and $\widehat{\beta}_1^*$ =b[3]+b[4]. So the fitted models are

$$ri_i = \begin{cases} 1.543203 - 0.000392 \times temp_i & \text{for Grzegorz} \\ 1.542842 - 0.000388 \times temp_i & \text{for Rachel} \end{cases}$$

5.7 Linear models for grouped data—One-way ANOVA

If you have read the last section then you will see the next section as a natural continuation of what we have discussed so far. **One-way analysis of variance**, often abbreviated **one-way ANOVA**, is a technique for analyzing grouped data. That is, we use ANOVA to look at data that has been measured on a single quantitative variable and grouped using a single categorical variable. The grouping variable divides the data into two or more groups. If there are only two groups in the data set, then we can analyze the data using the two-sample t-test. However, if we assume the variances are equal, then the two-sample t-test is exactly equivalent to

- one-way ANOVA
- a linear model with a single dummy variable

There are a number of equivalent ways to write the one-way ANOVA model. It is natural to think in terms of group means. If we have k groups (or k levels in our factor) then the model is

$$y_{ij} = \mu_i + \epsilon_{ij}, \quad \epsilon_{ij} \sim N(0, \sigma^2) \tag{5.5}$$

where y_{ij} is the jth measurement in the ith group, and μ_i is the mean of the ith group. The model simply says that every group has its own mean and a common standard deviation σ. Statisticians sometimes call this a **means model** for obvious reasons. A common alternative to the means model is the **effects model**. The effects model is expressed as

$$y_{ij} = \mu + \alpha_i + \epsilon_{ij}, \quad \epsilon_{ij} \sim N(0, \sigma^2) \tag{5.6}$$

where the effects, α_i are the "departures" from the grand mean μ in each of the groups, i.e.,

$$\alpha_i = \mu_i - \mu$$

The effects model makes sense in the context that we think of each of the levels of the factor having "an effect." For example, Mari et al. [39] were interested in the ante mortem urine concentration of gamma-hydroxybutyric acid (GHB). In their experiment, they had three groups of individuals: (1) volunteers with no alcoholic history, (2) alcoholics who were waiting to start a GHB treatment program but had not yet received the drug, and (3) alcoholics currently being treated with GHB. In this experiment, the experimenters were interested in the "effect" of the GHB supplement on the natural levels of GHB. The third formulation is a **regression model for ANOVA**. If a factor has k levels, then as we have seen the regression model will require $k-1$ dummy variables, so that equation 5.5 can be rewritten as

$$y_{ij} = \beta_0 + \beta_1 d_2 + \beta_2 d_3 + \cdots + \beta_{k-1} d_k + \epsilon_{ij}, \; \epsilon_{ij} \sim N(0, \sigma^2) \quad (5.7)$$

You might think this model looks unnecessarily complex, and perhaps it is. However, it is important for a number of reasons. Firstly, this parameterization highlights the fact that the strong distinctions that are often made between ANOVA models and regression models are artificial. Secondly, this is how R interprets and fits the ANOVA model. Whilst the mechanics of the fitting are not important to us, the interpretation of the model is. Therefore, it is important that we understand this parameterization so that we can correctly interpret the output from R.

If $i = 1$, then observation y_{ij} belongs to the "base" group, and all the dummy variables will be set to zero. In this case (5.7) will reduce to

$$y_{1j} = \beta_0 + \epsilon_{1j}$$

We can see from this equation, therefore, that $\mu_1 = \beta_0$. The "base" group is the bottom level of the factor. The choice of the base level is up to the experimenter. R defaults to an alphanumeric ordering of the levels if a particular ordering of the levels is not specified by the user. For example, if a factor has levels 1, 2, and 3, then by default R will choose level 1 as the base level. This seems reasonable. If, however, the levels are *Low*, *Medium*, and *High*, R will order them *High*, *Low* and *Medium* despite the fact that there is a natural ordering. Fortunately, it is easy to override the default, and we will discuss this in the tutorial section.

If $2 \leq i \leq k$, then the observation y_{ij} belongs to the ith group where i is anything except for the base. The ith dummy variable will have a value of one and all the others will be zero. Let us consider two examples where $i = 2$ and $i = 3$ to see how this works. If $i = 2$, then we know from our means model that

$$y_{2j} = \mu_2 + \epsilon_{2j}$$

and from our regression model we have

$$y_{2j} = \beta_0 + \beta_1 + \epsilon_{2j}$$

To make these two models equivalent, then we must have

$$\mu_2 = \beta_0 + \beta_1$$

and we know that $\beta_0 = \mu_1$ so

$$\beta_1 = \mu_1 - \mu_2$$

That is, β_1 quantifies the difference between the first and second levels of the factor. We can use exactly the same reasoning to show that when $i = 3$

$$\beta_2 = \mu_1 - \mu_3$$

In general, then

- $\beta_0 = \mu_1$. That is, the intercept β_0 is equal to the mean of the base (or first) level of the factor
- and $\beta_i = \mu_1 - \mu_i$, $i = 2, ..., (k-1)$

At this point, I hope you are asking "But what about all the other differences?" That is, the regression model shows us the difference between the means of every group and the first group, but not, for example, the difference between the second and third groups. This has two consequences. The first is that we cannot automatically declare that there are no differences between the any of the groups by examination of the P-values attached to the regression coefficients. The second is that we need to do a little bit of extra work if we want to examine all the pairs of differences. We can address the first issue by use of the ANOVA table. The solutions are more easily explained with an example.

5.7.1 Example 5.11—Differences mean RI in the same window

I have selected three panels from Bennett's data [9]. My reasons for choosing three panels instead of 49 will become clear later on Therefore, I have a set of 10 RI measurements from three different locations in a frame of glass. I wish to test the hypothesis that the true mean RIs of each of these locations are the same. That is,

$$H_0 : \mu_1 = \mu_2 = \mu_3$$

The alternative hypothesis is that "at least one" of these locations has a different mean. That is,

$$H_1 : \mu_i \ne \mu_j \text{ for some } i \ne j$$

Note that the alternative hypothesis says "for some" not "for all." That means that we do not require every group to be different from every other group to reject the null hypothesis. One approach to this problem would be to perform a two-sample t-test for every pair of groups. However, this is not advisable because of the problem of **multiple comparisons**.

The problem of multiple comparisons arises when we perform more than one hypothesis test to answer a question. The problem is that the Type I error compounds. That is, for each hypothesis test, we accept that there is a certain probability, α, that we will reject the null hypothesis even when it is true. This is a Type I error. If we perform multiple *independent* tests, then this probability is the same in each test. However, the chance we make "at least one" incorrect rejection in r tests is given by

$$Pr(\text{at least one Type I error}) = 1 - (1-\alpha)^r$$

If $\alpha = 0.05$ and $r = 1$, then this is 0.05. However, when $r = 2$, the probability almost doubles to $0.0975 \approx 0.1$, and it goes to 0.143 when $r = 3$. That is, with just three tests, we have almost tripled the chance of having a Type I error. One-way ANOVA (and ANOVA in general) provides a mechanism for avoiding Type I errors (at least initially) by simultaneously testing for all possible differences between the groups. In fact, we have been inadvertently relying on this. In multiple regression we look at the F-test for the significance of the regression. In this test, we simultaneously test the hypothesis that all of the regression coefficients are zero.

Returning to our example, we wish to see whether there are differences in the population means of our three panel locations. As always it is a good idea to plot the data if we can. We have 10 observations per group. Whilst it is technically possible to draw box plots for this data, it seems somewhat over the top to draw graphs that need five statistics to describe 10 data points. Figure 5.29 is a plot of the data by panel number. The arrows (generated using R's `arrows` command) are the means of each group plus or minus two standard errors. From this plot we can see that panel 7 has a substantially different sample mean from the other two panels. There are also three potential outliers that may (or may not) influence the result). One-way ANOVA lets us test the hypothesis that the true means of all three locations are the same. To perform this test, we use the `lm` command and the `anova` command. However, before we do so, in this circumstance we will rescale the data. The reason for this is that the numbers involved are very small. Small numbers can cause numerical instability or round-off error. R is quite robust to these problems, but it does have its limits. Most of the variation in R data from the same source occurs in the 4th to 6th decimal places. Therefore, we will subtract 1.5 from every RI and multiply the remainder by 1,000. That is,

$$\text{scaled } RI = 1000 \times (RI - 1.5)$$

This does have an effect on the data and some of the quantities involved, but it does not affect the F-statistic, the degrees of freedom, and the probability. As usual we examine the behavior of the residuals before we perform any inference. The residual plots we use are *exactly the same* as the ones we use for regression because the model is simply another linear model. The only difference is that the plot of the residuals against the fitted values will be quite patterned. The reason for this is that all of the observations in the same group have the same mean. Therefore, we will have vertical "lines" of points across the pred-res plot. Figure 5.30 shows the residual diagnostic plots. We can see that the residuals appear to have a slight left skew to them. This is apparent in the curvature of the normal Q-Q plot and the histogram of the residuals. However, this may be due to the presence of three large negative residuals. We are also in a position to make an additional check on the equality of variance assumption. We can gauge this by the relative spreads of the groups of points in the pred-res plot. Our judgment can be clouded a little by the presence of the outliers. If you hold your hand over these points, then the assumption of equal variance holds. We can now interpret the (one-way) ANOVA table for this data (Table 5.5). Many statistics concentrate on the relationships that

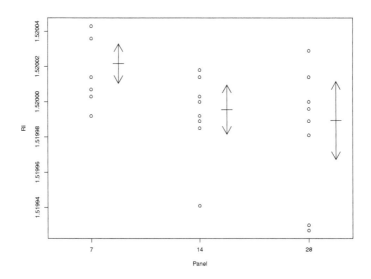

FIGURE 5.29: A plot of three different panels from the Bennett data set

	Df	Sum Sq	Mean Sq	F value	$\Pr(>F)$
panel	2	0.01	0.00	4.34	0.0233
Residuals	27	0.02	0.00		

TABLE 5.5: ANOVA table for the RI data

exist between elements in this table. However, the only one that I will highlight here is the degrees of freedom (Df) column. A k level factor always has $k-1$ degrees of freedom. The reason that this is important is because R must be told that a variable is a factor when the levels of the factor are numeric. If we do not do this, then R treats the model as a simple linear regression and the term in ANOVA table will have one degree of freedom. This is not incorrect, it is just not want we want to do. We can see from the output that our factor panel has two degrees of freedom, corresponding to the three levels, as desired. Another simple, useful check is that the Df column should sum to $n-1$, where n is the total number of observations. We had three groups of 10 observations, so $n = 30$, and we can see that $2 + 27 = 29 = 30 - 1$. This is a useful check because it can tell us if there are missing values or perhaps the data have not been loaded correctly. The key value in the ANOVA table is the P-value. If the P-value is large, then we can say that it appears the factor has no effect on the response, or that there is no evidence of a difference between the group means, and that is the end of our analysis. If the P-value is small, then we have evidence of an effect due to the factor, or evidence of a difference between the group means, and we proceed to look at the data further. In this case we can see that $0.01 < P = 0.0233 < 0.05$. Therefore, we have evidence of a difference between the group means. Our next question should be "Where are the differences?" This is an easier question to ask than it is to answer. We can examine some, but not all, of the differences by looking at the coefficients of the dummy variables in the regression table. We can see estimates of the

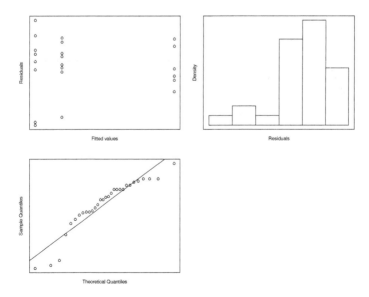

FIGURE 5.30: Residual diagnostic plots for a one-way ANOVA model

differences (and the standard error of the differences) between panels 7 and 14, and panels 7 and 28 from the regression table (Table 5.6) by looking at the regression coefficients for `panel14` and `panel28`. However, we cannot see an estimate or a standard error for the difference between panels 14 and 28. A simple way to deal with this when there are not many levels of the factor is to reorder the levels of the factor and refit the model. Remember the intercept provides an estimate of the mean of the base level, and all of the coefficients provide an estimate of the difference between the base level and the level of the factor given by the coefficient name. In this example the factor `panel` has levels 7, 14, and 28. By default R arranges the factor levels alphanumerically, so panel 7 is the base panel. To override this, we call the `as.factor` command and use the option to specify the order of the levels. E.g.,

```
> ri.df$panel = as.factor(ri.df$panel,
                          levels = c(14,28,7))
```

Now, when we fit the model, level 14 will be the base level, and we can get the comparisons with panels 14 and 28, and panels 14 and 7. Note that this "trick" works for the `lm` command, but it does not work for the `glm` we use in Chapter 6.

5.7.2 Three procedures for multiple comparisons

There are many procedures for dealing with multiple comparison issues. I will outline three common alternatives: **Bonferroni's correction**, **Fisher's (protected) LSD**, and **Tukey's HSD**.

5.7.2.1 Bonferroni's correction

Bonferroni's correction is the simplest of all the procedures to apply, and the most naïve. We carry out the F-test as normal, and if we fail to reject the null, then we do nothing. If we reject the null, then we either calculate a set of confidence intervals for the difference in each pair of means, or we perform two sample t-tests for the difference in each pair of means. The Bonferroni correction to this procedure is to divide our confidence level α value by the number of pairwise comparisons we are going to make. The reasoning is as follows. If the probability of making at least one Type I error in r comparisons is approximately $r \times \alpha$. Therefore, if we desire an overall Type I error rate of

	Estimate	Std. Error	t value	$\Pr(>|t|)$
(Intercept)	20.0217	0.0082	2432.44	0.0000
panel14	−0.0261	0.0116	−2.24	0.0334
panel28	−0.0323	0.0116	−2.77	0.0099

TABLE 5.6: Regression table for the RI one-way ANOVA model

α, we should set $\alpha_B = \alpha/r$ for each test, so that $r \times \alpha_B = \alpha$. This is a fairly reasonable solution if r is very small. The effect of using α_B instead of α is that it makes the confidence intervals wider (and hence more likely to include zero), and it is harder to reject the null hypothesis because the rejection region is smaller.

The t-tests for the difference between the two group means can be done with the *pooled* t-test, by setting var.equal = TRUE in t.test. This is consistent with the assumption that the data have common variance regardless of group. Alternatively, the assumption of equal variance can be dropped and *Welch's* t-test can be performed. This is the default option for t.test. The confidence intervals are produced as part of the t.test output. The Bonferroni correction is made by setting the conf.level parameter in the t.test call to $1 - \alpha_B$.

In our example the statistics are

Diff.	t-stat	Pooled P	Pooled CI	Welch P	Welch CI
$\mu_7 - \mu_{14}$	2.91	0.0094	(0.0024, 0.0498)	0.0098	(0.0023, 0.0499)
$\mu_7 - \mu_{28}$	2.60	0.0180	(-0.0005, 0.0651)	0.0216	(-0.0017, 0.0663)
$\mu_{14} - \mu_{28}$	0.47	0.6418	(-0.0284, 0.0408)	0.6428	(-0.0290, 0.0414)

Note that the test statistics are equal for the pooled test and the Welch when the sample sizes are equal. The P-values and confidence intervals will be very slightly different because the degrees of freedom are different between the two tests.

5.7.2.2 Fisher's protected least significant difference (LSD)

Fisher's (protected) LSD is a solution suggested by R. A. Fisher. We carry out the F-test as normal, and if we fail to reject the null, then we do nothing. If we reject the null, then we compare the difference in each pair of means to

$$LSD = t_\nu(\alpha/2)\hat{\sigma}\sqrt{\frac{1}{n_A} + \frac{1}{n_B}} \qquad (5.8)$$

where

- $\hat{\sigma}$ is the residual standard error from the regression table (or the square root of the residual mean square from the ANOVA table)

- n_A and n_B are the sample sizes of the two groups being compared

- ν is the degrees of freedom association with the residual standard error (or the residual degrees of freedom from the ANOVA table)

The "protected" part of the name comes of the name relates to the performance of the initial F-test. It was originally thought that this initial step controlled the Type I error rate.

We can also adjust this value with the Bonferroni correction. This procedure is sometimes called "Fisher's protected LSD with a Bonferroni correction." We compare the difference in each pair of means with

$$LSD_B = t_\nu(\alpha_B/2)\hat{\sigma}\sqrt{\frac{1}{n_A} + \frac{1}{n_B}} \tag{5.9}$$

In our example $n_A = n_B = 10$, $\hat{\sigma} = 0.0260$, and $\nu = 30 - 3 = 27$. If $\alpha = 0.05$, then we compare the absolute value of the difference between each pair of means with

$$LSD = t_{27}(0.05) \times 0.0260 \times \sqrt{\frac{1}{10} + \frac{1}{10}}$$
$$= 0.0239$$

If we choose to apply the Bonferroni correction as well, then $\alpha_B \approx 0.0167$, and we compare the absolute value of the difference between each pair of means with

$$LSD_B = t_{27}(0.0167) \times 0.0260 \times \sqrt{\frac{1}{10} + \frac{1}{10}}$$
$$= 0.0297$$

If we require confidence intervals rather than significance tests we can obtain these from

$$\bar{y}_i - \bar{y}_j \pm LSD \tag{5.10}$$

where LSD_B can be substituted for LSD if desired.

5.7.2.3 Tukey's Honestly Significant Difference (HSD) or the Tukey-Kramer method

Tukey's HSD is named after the famous statistician John Tukey who invented it. The extension of the method to unequal group sizes is attributed to statistician Clyde Kramer. The procedure uses the **Studentized range distribution**, which is a probability distribution for the range (the difference between the maximum and the minimum) of a sample of size n from a normal distribution with standard deviation σ. It is generally regarded as being more conservative than Fisher's LSD but less conservative than Bonferroni's correction. Conservative in this setting means that it "admits more uncertainty in the estimates" by producing wider confidence intervals for the differences. The difference in each pair of means is compared to

$$TSR = \frac{q_{1-\alpha;k,N-k}}{\sqrt{2}}\hat{\sigma}\sqrt{\frac{1}{n_A} + \frac{1}{n_B}} \tag{5.11}$$

where

- $q_{1-\alpha;k,N-k}$ is the $1-\alpha$ quantile from the Studentized range distribution. This can be obtained from the `qtukey` function in R.
- r is the number of groups or levels of the factor
- $N-r$ is the degrees of freedom of the residual standard error (or the total number of observations minus the number of groups)
- $\hat{\sigma}$ is the residual standard error from the regression table (or the square root of the residual mean square from the ANOVA table)
- and n_A and n_B are the sample sizes of the two groups being compared

When $n_A = n_B = n$, TSR reduces to

$$TSR = q_{1-\alpha;k,N-k} \frac{\hat{\sigma}}{\sqrt{n}} \tag{5.12}$$

If confidence intervals are required rather than significance tests then these can be obtained for each pair of means from

$$\bar{y}_i - \bar{y}_j \pm TSR$$

In our example $n = 10$, $N = 30$, $k = 3$, and $\hat{\sigma} = 0.0260$ so we compare each pair of means to

$$TSR = q_{0.95;3,27} \frac{0.0260}{\sqrt{10}}$$
$$= 3.5064 \times 0.0082$$
$$= 0.0289$$

5.7.2.4 Which method?

We have seen three different methods. A natural question to ask is, "Which one is best?" As a general rule of thumb, the methods, in order of highest to lowest conservativeness are Bonferroni, LSD_B, TSR, LSD. This means on average, Bonferroni will produce the widest confidence intervals, followed by LSD_B, TSR, and then LSD. Wider intervals mean that we decrease our probability of making a Type I error, but we increase our probability of making a **Type II error**. A Type II error occurs when we accept the null hypothesis even though the alternative is actually true. That is, we fail to detect a difference when it is present. The Type I error rate of a statistical test is called the **size** of the test. The Type II error rate is sometimes denoted β. $1-\beta$ or one minus the Type II error rate is called the **power** of a test. There is always a trade-off between size and power. If we optimize our test so that

Difference	Bonferroni	LSD_B (0.0297)	TSR (0.0289)	LSD (0.0239)
$\mu_7 - \mu_{14} = 0.026$	+	−	−	+
$\mu_7 - \mu_{28} = 0.032$	−	+	+	+
$\mu_{14} - \mu_{28} = 0.006$	−	−	−	−

TABLE 5.7: Results of (two-tailed) significance tests for a difference between the two means. A + means that the test was significant at the 0.05 level or that the 95% confidence interval did not contain zero

we minimize size, we do this at a cost of decreasing power. That is, it becomes harder to detect a difference. If we optimize a test so that we maximize our chance of finding a difference, i.e., increase power, then the size of the test will increase. That is, it becomes more likely that we will say two things are different when we are not. Therefore, "which method" is a balance between what we are willing to accept in terms of size and power of the procedure.

In this example, the interval widths for the Bonferroni, LSD_B, TSR, LSD methods are 0.0620, 0.0594, 0.0577, and 0.0478, respectively. Note that the value for the Bonferroni is an average for the widths of the confidence intervals based on the Welch correction.

Table 5.7 shows the results of the (two-tailed) significance tests for a difference between the two means. A + means that the test was significant at the 0.05 experiment-wise level. By experiment-wise I mean that we have attempted to control the Type I error rate so that it is 0.05 for all three tests. In the case of the Bonferroni method and the LSD Bonferroni method, this means using $\alpha_B = 0.05/3 \approx 0.0167$. Based on these results we would probably say that there is definitely a difference between panels 7 and 28, and possibly between panels 7 and 14, although the evidence is weaker for this. Panels 14 and 28 are much more similar to each other than to panel 7. In fact we can test this idea explicitly using a **linear contrast**.

5.7.2.5 Linear contrasts

Linear contrasts provide a framework for formulating more general hypotheses. For example, we can test the hypothesis that the mean of panel 7 the same as the mean of panels 14 and 28, i.e.,

$$H_0 : \mu_7 - \frac{1}{2}(\mu_{14} + \mu_{28}) = 0$$

In general, a linear contrast can be used to test the hypothesis that

$$H_0 : \sum_{i=1}^{k} c_i \mu_i = 0$$

where $\sum c_i = 0$. We have unknowingly been using this method all along, because all the pairwise differences between groups i and j can be written as a contrast with, $c_i = 1$, $c_j = -1$ and $c_l = 0$ for all $1 \leq l \neq i, j \leq k$. For example, with the glass data from panes 7, 14, and 28, three linear contrasts are

Difference	c_7	c_{14}	c_{28}
$\mu_7 - \mu_{14}$	1	-1	0
$\mu_7 - \mu_{28}$	1	0	-1
$\mu_{14} - \mu_{28}$	0	1	-1

It is easy to see that the sum of the coefficients in each row is zero. We can test the hypotheses associated with each of these contrasts by calculating

$$\widehat{\theta} = \sum_i^k c_i \bar{y}_i \tag{5.13}$$

where $\widehat{\theta}$ is the sample estimate of the contrast θ and comparing it (in absolute value) to

$$LSD = t_{N-k}(\alpha/2) \times se\left(\widehat{\theta}\right)$$

$$= t_{N-k}(\alpha/2) \times \widehat{\sigma} \sqrt{\sum_i^k \left(\frac{c_i}{n_i}\right)^2} \tag{5.14}$$

where $\widehat{\sigma}$ is the residual standard error for the model. We can see that when we are comparing pairs of means, equation (5.14) is equal to equation (5.8) as it should be because this is a general framework which can be used to do what we have already done.

Therefore, for our specific problem, we have $c_1 = 1$, $c_{14} = c_{28} = -0.5$, so

$$\widehat{\theta} = 1 \times \bar{y}_7 - 0.5 \times \bar{y}_{14} - 0.5 \times \bar{y}_{28}$$
$$= 0.0292$$

and we compare this to

$$LSD = t_{27}(0.025) \times \widehat{\sigma} \sqrt{\left(\frac{1}{10}\right)^2 + \left(\frac{-0.5}{10}\right)^2 + \left(\frac{-0.5}{10}\right)^2}$$
$$= 0.0065$$

$\widehat{\theta}$ is obviously much greater than LSD; hence, there is very strong evidence that there is a difference between the mean of panel 7 and the mean of panels 14 and 28. We can get the exact P-value by calculating a test statistic

$$T_0 = \frac{\widehat{\theta}}{se\left(\widehat{\theta}\right)} = \frac{\sum_{i=1}^k c_i \bar{y}_i}{\widehat{\sigma} \sqrt{\sum_{i=1}^k \left(\frac{c_i}{n_i}\right)^2}} \tag{5.15}$$

and then finding $Pr(T \geq |T_0|)$ using a Student's t distribution with $N - k$ degrees of freedom. For our example $T_0 = 9.16$, which gives a P-value of effectively zero since it is over nine standard deviations above the mean.

5.7.3 Dropping the assumption of equal variances

In many situations, the assumption of equal variance between the groups is not realistic. Wild and Seber [16] say that the F-test is reasonably robust to departures from the assumption of equal variances. As a conservative rule of thumb they suggest that if the ratio of the largest variance to the smallest variance is less than four (or two if you are using standard deviation), and the group sizes are roughly equal, then the F-test will perform adequately. The confidence intervals, however, are not robust. This is because the equality of variance assumption means that all of the confidence intervals will be the same width if the sample sizes are equal and approximately the same width if the sample sizes are unequal. If there are truly differences in the different groups, then the intervals will be too wide for the cases where the variability is low and too narrow for the cases where the variability is high.

We have seen how we can remedy this occasionally by transforming the response. This approach does not always work, or does not completely remove the problem.

An alternative is to assume that the variance of each observation or variance of residual, if you prefer, in each group is of the form $w_i \sigma^2$, where w_i is a **weight** that is specific to the ith observation. This weight allows the variance to change on an observation by observation basis. There is a catch. To paraphrase Miller [40] who so beautifully puts it "If God or the Devil were willing to tell us the weights, then there is a simple procedure that allows us to include them in our estimation process. Since most of us cannot get help from either above or below, we are faced with having to estimate the weights." The procedure that incorporates the weights is called **weighted least squares** or **WLS** for short. We have been using **ordinary least squares**, or **OLS**, up until now. As noted, the problem is that we do not generally know the values of the weights. However, in the one-way ANOVA case, we do have a reasonable method of estimating them. The weight of every observation in the ith group is set to

$$\frac{1}{s_i^2}$$

where s_i^2 is the sample variance for the ith group, respectively. The procedure is known sometimes as the Welch-James correction [41, 42]. To see how this procedure works and how we incorporate the weights, we will look at an example.

5.7.3.1 Example 5.12—GHB concentration in urine

Mari et al. [39] were interested in determining the ante mortem urine concentration of gamma-hydroxybutyric (GHB) acid. GHB is produced naturally, but it is also used as a medication for a variety of clinical conditions and can be abused by recreational users. According to Mari et al. the important issue for forensic scientists is whether the detection of GHB in urine, in concentrations above some yet to be determined value, can be used as evidence for drug facilitated assault. In an attempt to set a cut-off level, Mari et al. took urine sample from three sets of people:

A. 30 alcoholics who are being treated with known oral doses of GHB (marketed commercially as Alcover®) as a treatment of alcohol withdrawal and dependence,

B. 30 volunteers who had no exogenous GHB intake, and

C. 30 alcoholics before they had initiated GHB therapy,

and measured the endogenous GHB concentration (μg/mL) in each. The disparity in the variability in each group is obvious in (left-hand side of Figure 5.31). The problem is improved but not removed when we take logarithms (right-hand side of Figure 5.31), which is a standard transformation for concentration data. The ratio of the largest to smallest variance is 14.2, which

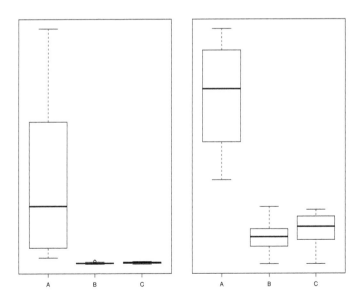

FIGURE 5.31: Gamma-hydroxybutyric (GHB) acid concentration in three groups of people

182 Introduction to data analysis with R for forensic scientists

is well in excess of four! Therefore, we might consider using WLS. There will be no difference between the estimated group means in this example using either WLS or OLS to fit the model. This is because the experiment has equal numbers of observations per treatment group. When this is the case the experiment is said to be **balanced**. The difference will be apparent, however, in the confidence intervals for the group means which can be seen in Figure 5.32. We can see that from Figure 5.32 that the intervals behave as we would like them to in that they are wider for group A and narrower for groups B and C.

5.7.3.2 An alternative procedure for estimating the weights

There is an alternative procedure for estimating the weights, which is sometimes called **residualizing the x's**. The procedure is iterative and continues until the regression coefficients converge. **Convergence** means here that there is minimal to no change in the regression coefficients in subsequent steps.

1. Set $w_i = 1$ for each observation

2. Repeat the following steps until the regression coefficients converge

    i. Regress y_i against x_i with weights w_i

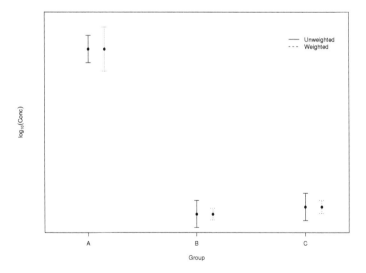

FIGURE 5.32: 95% confidence intervals for the (log) GHB data using unweighted and weighted least squares

The linear model

Panel	Welch-James	Residualized
7	3171.1	3523.5
14	2035.1	2261.2
28	815.8	906.4

TABLE 5.8: Comparison of Welch-James and residualized weights

ii. Compute the residuals, r_i, from the fitted model in the previous step

iii. Regress r_i^2 against x_i

iv. Let the new weights be $w_i = 1/\widehat{r}_i^2$ where the $\widehat{r}_i^2$'s are the fitted values from the previous step

This method works by making the new weights inversely proportional to the observations that have large residuals, i.e., have a large amount of variation. This is analogous to the Welch-James method, which assigns weights that are inversely proportional to the group variances. Reassuringly this iterative scheme produces almost identical weights to the Welch-James weights in the balanced one-way ANOVA case.

5.7.3.3 Example 5.13—Weighted least squares

The ratio of the largest to smallest variances for the panel RI data is 3.9, which is on the border of our threshold. As a point of illustration, we will use the iterative scheme to generate the group weights and compare them to the Welch-James weights. The iterative procedure was run for 10 iterations and the generated weights and the Welch-James weights are shown in Table 5.8. The weights are approximately the same, and the confidence intervals for the groups are identical using either method.

5.7.4 Tutorial

1. Load the Bennett et al. data.

    ```
    > data(bennett.df)
    ```

2. Make a new data frame that has two columns. The first column is called **ri** and holds the data from panels 7, 14, and 28. The second column (called **panel**) is a factor that tells R which panel each observation belongs to.

    ```
    > panel.df = data.frame(
                ri = with(bennett.df, c(X7,X14,X28)),
                panel = factor(rep(c(7,14,28),rep(10,3))))
    ```

3. Add a column to the data called `ri.scaled`, which is the `ri` value for each observation scaled by subtracting 1.5 and multiplying by 1,000.

   ```
   > panel.df = within(panel.df, {
        ri.scaled = 1000 * (ri - 1.5)
   })
   ```

4. Draw a box plot of the scaled RI by panel.

   ```
   > boxplot(ri.scaled ~ panel, data = panel.df)
   ```

5. Use the `onewayPlot` from the `s20x` library to plot the raw observations by group.

   ```
   > library(s20x)
   > onewayPlot(ri.scaled ~ panel, data = panel.df)
   ```

   Note the *outlying* points in the last two plots. We will come back to these values.

6. Fit a one-way ANOVA model to this data with `ri.scaled` as the response and `panel` as the factor variable.

   ```
   > panel.fit = lm(ri.scaled ~ panel, data = panel.df)
   ```

7. Check out the residual diagnostic plots.

   ```
   > plot(panel.fit, which = 1:2)
   ```

8. Have a look at the ANOVA table for the model to see if `panel` has "an effect" on `ri.scaled`. That is, see if there is any evidence of a difference between any of the groups.

   ```
   > anova(panel.fit)

   Analysis of Variance Table

   Response: ri.scaled
             Df  Sum Sq   Mean Sq  F value  Pr(>F)
   panel      2  0.00588  0.002938    4.34   0.023 *
   Residuals 27  0.01829  0.000678
   ---
   Signif. codes:  0 '***' 0.001 '**' 0.01 '*' 0.05 '.' 0.1 ' ' 1
   ```

9. An easy way to perform Tukey's HSD (or to get HSD intervals) is to use the `multipleComp` command from the `s20x` library. I am reluctant to use too many external R packages because usually I have no control over them. The problem with that is that I cannot guarantee that they will always work in the same way. However, in this instance, I am the maintainer of the `s20x` library so it should stay fairly stable. The `multipleComp` function only works for one-way ANOVA models.

The linear model   185

```
> multipleComp(panel.fit)

         Estimate Tukey.L Tukey.U Tukey.p
7  - 14    0.0261 -0.0028  0.0550  0.0820
7  - 28    0.0323  0.0034  0.0612  0.0259
14 - 28    0.0062 -0.0227  0.0351  0.8561
```

10. One question we might have is, "Are the outliers driving the significance?" That is, "Are the extreme low values (one in panel 14 and two in panel 28) the reason that there is a significant difference?" We can test this idea by refitting the model without these values. To do this we need to identify them. An easy way to this is to plot the observations by panel number, by using the observation number as the plotting symbol.

```
> x = rep(1:3, rep(10, 3))
> plot(ri.scaled ~ x, type = "n", data = panel.df)
> text(x, panel.df$ri.scaled, 1:30)
```

We use have to use the `text` command here, because using `pch` with will only plot one character. Using this method we can see (Figure 5.33) our extreme points are observations 17, 26, and 30. We can refit the model without these points using the `subs` option of the `lm` command. The `subs` option allows us to specify:

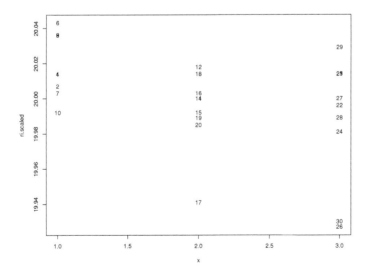

FIGURE 5.33: The (scaled) RI values plotted by group with observation numbers used as the plotting character

i. a vector of observation indices we wish to include

ii. a vector of observation indices we wish to exclude (by making them negative or FALSE)

iii. a vector of logical values where TRUE means include the observation and FALSE means exclude the observation.

We will use the second form

```
> panel.fit2 = lm(ri.scaled~panel,
                subs = -c(17,26,30), data = panel.df)
> anova(panel.fit2)

Analysis of Variance Table

Response: ri.scaled
          Df   Sum Sq  Mean Sq F value Pr(>F)
panel      2 0.00222 0.001112    4.68  0.019 *
Residuals 24 0.00570 0.000238
---
Signif. codes:  0 '***' 0.001 '**' 0.01 '*' 0.05 '.' 0.1 ' ' 1

> multipleComp(panel.fit2)

            Estimate  Tukey.L  Tukey.U  Tukey.p
7  - 14    0.0200333   0.0023   0.0377   0.0243
7  - 28    0.0170750  -0.0012   0.0353   0.0699
14 - 28   -0.0029583  -0.0217   0.0157   0.9179
```

The answer to our question is an interesting "Yes and no." We can see in the ANOVA table that the P-value actually decreased. In the original model, the only significant difference was between panels 7 and 28. In the refitted model, the only significant difference is between panels 7 and 14. The evidence for a difference between 7 and 28 is on the weak side.

11. Load the GHB data.

    ```
    > data(ghb.df)
    > names(ghb.df)
    ```

    [1] "sample" "group1" "group2" "group3"

12. Reorganize the data so that the response, conc is in a single vector and the groups are labeled A, B, and C in a single vector group.

    ```
    > ghb.df = data.frame(conc = unlist(ghb.df[,-1]),
        group = factor(rep(LETTERS[1:3],rep(30,3))))
    > row.names(ghb.df) = paste(ghb.df$group,
              rep(1:30,3),sep="-")
    ```

The linear model

The built in vector `LETTERS` contains all the uppercase letters from A–Z. Correspondingly, `letters` contains all the lowercase letters. You might also find `month.names` and `month.abb` useful.

13. Produce a box plot of the GHB concentration by group. Try the logarithm base 10 and the inverse square root $(-1/\sqrt{conc})$ as well.

```
> par(mfrow = c(2, 2))
> boxplot(conc ~ group, data = ghb.df)
> boxplot(log10(conc) ~ group, data = ghb.df)
> boxplot(-1/sqrt(conc) ~ group, data = ghb.df)
```

Note that we use $-1/\sqrt{conc}$, instead of $1/\sqrt{conc}$ just to preserve the order of the data. That is, if we do not change the sign, then the largest values become the smallest and vice versa.

14. Create two new variables in the data frame, `log10.conc` and `inv.sqrt.conc`

```
> ghb.df = within(ghb.df, {
      log10.conc = log10(conc)
      inv.sqrt.conc = -1/sqrt(conc)
  })
```

15. Fit four one-way ANOVA models to the concentration data

Model 1: `conc` grouped by `group`

Model 2: `log10.conc` grouped by `group`

Model 3: `log10.conc` grouped by `group` and weighted by the inverse of the group sample variances

Model 4: `inv.sqrt.conc` grouped by `group`

```
> ghb.fit1 = lm(conc ~ group, data = ghb.df)
> ghb.fit2 = lm(log10.conc ~ group, data = ghb.df)
> ghb.vars = with(ghb.df, sapply(split(log10.conc,
      group), var))
> ghb.weights = rep(1/ghb.vars, rep(30, 3))
> ghb.fit3 = lm(log10.conc ~ group, weights = ghb.weights,
      data = ghb.df)
> ghb.fit4 = lm(inv.sqrt.conc ~ group, data = ghb.df)
```

16. Look at the diagnostic plots for each of the four models.

```
> par(mfrow = c(2, 2))
> plot(ghb.fit1)
> plot(ghb.fit2)
> plot(ghb.fit3)
> plot(ghb.fit4)
```

17. Look at the ANOVA tables and summary tables for models 2–4.

    ```
    > anova(ghb.fit2)
    > summary(ghb.fit2)
    > anova(ghb.fit3)
    > summary(ghb.fit3)
    > anova(ghb.fit4)
    > summary(ghb.fit4)
    ```

18. Get prediction intervals for a concentration of a new person using models 2–4 and back transform them to their original scale. We get prediction intervals for a new person, rather than confidence intervals for the mean because the original authors were interested in setting a threshold value where one might declare that a victim had been given a dose of GHB. A confidence interval for the mean would tell us the range of values we might expect to see if we took measurements from 30 people. We can use **predict** to calculate prediction intervals for models 2 and 4, but not for model 3. If we use a weighted model for prediction intervals, R will give us a warning that a constant estimate for variance is used. The correct intervals are given by

    $$\hat{\mu}_i + t(\alpha/2; N-k)\sqrt{se(\hat{\mu})^2 + \frac{Res\ MS}{w_i}}$$

    In the one-way ANOVA case, the Residual Mean square is 1, so

    $$\frac{Res\ MS}{w_i} = s_i^2$$

    The standard error of the fitted mean can be obtained from predict using the se = TRUE option.

    ```
    > ghb.pred2 = 10^predict(ghb.fit2,
                      data.frame(group = LETTERS[1:3]),
                      interval = "prediction")
    > mu  = predict(ghb.fit3, data.frame(group = LETTERS[1:3]),
                se = T)
    > lb = 10^(mu$fit-qt(0.975,87)*sqrt(mu$se^2+ghb.vars))
    > ub = 10^(mu$fit+qt(0.975,87)*sqrt(mu$se^2+ghb.vars))
    > ghb.pred3 = cbind(10^mu$fit, lb, ub)
    > row.names(ghb.pred3) = LETTERS[1:3]
    > ghb.pred4 = 1/predict(ghb.fit4,
                      data.frame(group = LETTERS[1:3]),
                      interval = "prediction")^2
    > ghb.pred2
            fit     lwr     upr
    1 18.17855 4.25718 77.6241
    ```

```
2  0.77028 0.18039  3.2891
3  0.88417 0.20706  3.7755
```

```
> ghb.pred3
```

```
         lb              ub
A  18.17855 1.75894 187.8743
B   0.77028 0.41416   1.4326
C   0.88417 0.44158   1.7704
```

```
> ghb.pred4
```

```
       fit     lwr      upr
1 13.14044 2.49233 149.8140
2  0.75305 0.43862   1.5830
3  0.85876 0.48449   1.9208
```

19. (**Advanced**) Mari et al. reported that other authors had suggested a threshold concentration of 10 μg/mL. In choosing such a threshold, we would like the probability of someone who had actually had an artificial dose of GHB falling below the threshold to be very low, and the probability of someone who had not had any GHB falling above it to be very high. The prediction intervals for our data are very tight for groups B and C, and so we can deduce that the chance of someone who has not had any external dose of GHB having a concentration above 10 μg/mL is very low. Therefore, we proceed by setting the threshold with respect to group A, and then we evaluate the risk for groups B and C.

   (a) Firstly, we find and store the fitted means for group A and the standard error of the fitted means using models 2–4.

   ```
   > mu = list(mod2 = predict(ghb.fit2,
                              data.frame(group = "A"),
                              se = T),
               mod3 = predict(ghb.fit3,
                              data.frame(group = "A"),
                              se = T),
               mod4 = predict(ghb.fit4,
                              data.frame(group = "A"),
                              se = T))
   ```

   (b) Next, we allocate space to store the fitted probabilities and thresholds for models 2–4.

   ```
   > probs = list(mod2 = rep(0, 200), mod3 = rep(0, 200),
         mod4 = rep(0, 200))
   > threshold = seq(2, 10, length = 200)
   > thresh.05 = rep(0, 3)
   > thresh.10 = rep(0, 3)
   ```

(c) Now we standardize each of the threshold values by the mean and standard deviation from each model, and calculate the lower tail probability (i.e., the probability of falling below that threshold) using a t-distribution with 87 degrees of freedom.

```
> for(i in 2:4){
    listName = paste("mod", i, sep = "")
    pred = mu[[listName]]
    mx = pred$fit
    if(i!=3){
        sx = with(pred,sqrt(se.fit^2+
                 residual.scale^2))
    }else{
        sx = with(pred,sqrt(se.fit^2+ghb.vars[1]))
    }

    z = threshold
    if(i<4){
        z = (log10(threshold)-mx)/sx
    }else{
        z = (-1/sqrt(threshold)-mx)/sx
    }
    probs[[listName]] = pt(z,87)
}
```

(d) Finally, we plot the resulting probability curves.

```
> plot(threshold, probs[["mod4"]], type = "l",
    xlab = "Threshold", ylab = "Probability")
> lines(threshold, probs[["mod3"]], lty = 2)
> lines(threshold, probs[["mod2"]], lty = 3)
> legend(2, 0.4, lty = 3:1,
        legend = c("Model 2", "Model 3", "Model 4"),
        bty = "n")
> abline(h = 0.05)
```

We can see from Figure 5.34 that a threshold of 10 μg/mL has an unacceptably high chance of failing to include a person who truly has ingested GHB.

(e) We can set a threshold by choosing an *acceptable level of risk*. For example, if we choose 0.1, then we say that "We accept there is about a 10% probability that we will say that someone has not ingested GHB when they truly have." We can use the approx command to interpolate the threshold for a given probability for each model,

```
> thresh.01 =
    c(approx(probs[["mod2"]],threshold,0.1)$y,
```

The linear model

```
              approx(probs[["mod3"]],threshold,0.1)$y,
              approx(probs[["mod4"]],threshold,0.1)$y)
> thresh.01
[1] 7.0788 3.9862 3.8723
```

and then we can evaluate the risk for people who have not ingested GHB of being misclassified as having done so. The code here does this for people from group B. It can be changed to group C by changing the second line to group = "C".

```
> probs.01 = rep(0,3)
> group = "B"
> mu = list(mod2 = predict(ghb.fit2,
                      data.frame(group = group),
                      se = T),
            mod3 = predict(ghb.fit3,
                      data.frame(group = group),
                      se = T),
            mod4 = predict(ghb.fit4,
                      data.frame(group = group),
                      se = T))
> for(i in 2:4){
    listName = paste("mod",i,sep="")
    pred = mu[[listName]]
```

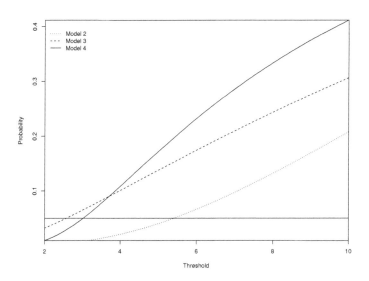

FIGURE 5.34: The probability that GHB measurement would fall below a threshold given that the person had received GHB

```
        mx = pred$fit
        if(i!=3){
            sx = with(pred,sqrt(se.fit^2
                                    +residual.scale^2))
        }else{
            sx = with(pred,sqrt(se.fit^2+ghb.vars[1]))
        }

        z = 0
        if(i<4){
            z = (log10(thresh.01[i-1])-mx)/sx
        }else{
            z = (-1/sqrt(thresh.01[i-1])-mx)/sx
        }
        probs.01[i-1] = 1-pt(z,87)
    }
> probs.01

[1] 0.00157618 0.07953031 0.00028165
```

We can see from this exercise that if we use the WLS model and a threshold of 3.99 μg/mL we have about an 8% chance of misclassifying a case. This rises to about 10% in group C. If we use the inverse square root model, which on the whole seems to perform better, a threshold of 3.87 μg/mL has about 0.02% chance of being misclassified (and this rises to 1% in group C).

(f) Finally, we can see how well our thresholds work for the observed data.

```
> table(ghb.df$group[ghb.df$conc < thresh.01[3]])

 A  B  C
 4 30 30
```

Using 3.87 as our cut-off point we can see from the output above that we would misclassify 4 out of our 30 *cases* (people who have actually ingested) but none of our *controls*. **Note:** You should treat this analysis with a little skepticism. Intervals based on data that have been weighted and/or transformed have biases in them that can be difficult to remove. There are remedies, see Carroll and Ruppert [43], for example, but they are beyond the scope of this book. Having said that, apart from the upper bounds of the prediction intervals being a little high, I am quite happy with the thresholds.

5.8 Two-way ANOVA

We have seen how we can go from a simple linear regression model, to a multiple regression model, to a multiple regression model for one-way ANOVA. Therefore, a logical extension of the one way ANOVA model is the addition of one or more factors. When the model contains two factor variables and no continuous variables, it becomes a **two-way analysis of variance** model or a **two-way ANOVA** model. In fact we can have any number of factors in our model for an n **way ANOVA**. The limiting feature, however, is usually the availability of data. That is, unless we use specialized experimental designs or make some compromises, we need at least two observations for every combination of the levels of the factors. We will return to this idea of **replication** later.

The two-way ANOVA model, as we have explained, contains two factors with I levels in the first and J levels in the second. These two factors are used to divide our observations into $I \times J$ groups. At some level, we are interested in whether there are differences between these groups. The reason that this statement is not whole-heartedly committed to this idea is because of the ways that we might think about how the response is affected by the different levels of factors. There are two-ways. Either the effects of the different factor levels are additive, or they interact. It sometimes helps to think of interaction as multiplicative. A simple example may explain the concept of interaction. Imagine an experiment where we are interested in the corrosive effects of tap water and salt (sodium chloride). We devise an experiment where we have two levels for each of our factors:

- $\overline{W}$ = no water, W = water

- $\overline{S}$ = no salt, S = salt

We then allocate our experimental units (strips of iron) to one of the four groups formed by the combination of these factors: $\{\overline{WS}, W\overline{S}, \overline{W}S, WS\}$. When we carry out this experiment, we expect

- A minimal to non-existent amount of corrosion when nothing is applied to the experimental units ($\overline{WS}$).

- A small amount of corrosion when water but no salt is applied to the experimental units ($W\overline{S}$).

- A small amount of corrosion when dry salt but no water is applied to the experimental units ($\overline{W}S$).

- A large amount of corrosion when water and salt are combined and applied to the experimental units.

That is, we expect water and salt to interact. The behavior is demonstrated graphically by the non-parallel lines in the **interaction plot** on the left of Figure 5.35. When there is no interaction (right of Figure 5.35), then we expect the amount of corrosion to increase by the same amount with the addition of salt, regardless of whether water is present. This is represented by the parallel lines in the interaction plot.

The two-way ANOVA means model is

$$y_{ijk} = \mu_{ij} + \epsilon_{ijk}, \ \epsilon_{ijk} \sim N(0, \sigma^2)$$

That is, we believe each group of observations, defined by having level i of factor A, and level j of factor B has the same mean μ_{ij}. Note that exactly the same assumptions about the residuals that have been made for every other linear model are also made here. The two-way ANOVA effects model can be more useful because it explicitly defines the interaction

$$y_{ijk} = \mu + \alpha_i + \beta_j + (\alpha\beta)_{ij}, \ \epsilon_{ijk} \sim N(0, \sigma^2)$$

The term α_i reflects the effect due to the ith level of factor A, and the term β_j reflects the effect due to the jth level of factor B. The term $(\alpha\beta)_{ij}$ represents the effect due the interactions of factors A and B. Of course, there may very well be no interactions between the factors, in which case this term can be dropped from the model.

The regression model for two-way ANOVA initially looks complicated. If

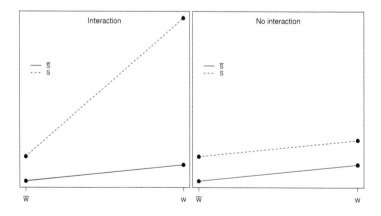

FIGURE 5.35: An example of interaction

factor A has I levels, and factor B has J levels then we need $I-1$ dummy variables for factor A and $J-1$ dummy variables for factor B. The interactions are represented by multiplying the respective dummy variables for level i of factor A and level j of factor B together. This can be seen if we return to our water/salt experiment. Each of the factors has two levels. Therefore, the regression model for this experiment is

$$y_{ijk} = \beta_0 + \beta_1 W_{ijk} + \beta_2 S_{ijk} + \beta_3 (W_{ijk} \times S_{ijk}) + \epsilon_{ijk} \quad \epsilon_{ijk} \sim N(0, \sigma^2)$$

where

- $W_{ijk} = 1$ if the ijk^{th} unit has received water and 0 otherwise

- $S_{ijk} = 1$ if the ijk^{th} unit has received salt and 0 otherwise

Using this parameterization we can see that when no water or salt are added $W_{ijk} = S_{ijk} = 0$, and

$$\beta_0 = \mu_{\overline{W}\overline{S}}$$

Similarly, when water is added but salt is not then $W_{ijk} = 1, S_{ijk} = 0$, and

$$\mu_{W\overline{S}} = \beta_0 + \beta_1$$
$$= \mu_{\overline{W}\overline{S}} + \beta_1$$
$$\beta_1 = \mu_{W\overline{S}} - \mu_{\overline{W}\overline{S}}$$

and $\mu_{W\overline{S}} = \beta_0 + \beta_1$. We can use the same reasoning to show that

$$\beta_2 = \mu_{\overline{W}S} - \mu_{\overline{W}\overline{S}}$$
$$\beta_3 = \mu_{WS} - \mu_{\overline{W}\overline{S}}$$

In this special situation where each factor only has two levels every single group mean can be estimated from the regression parameters. This is not usually the case. However, estimating the group means is not difficult and we will cover this in the tutorial.

5.8.1 The hypotheses for two-way ANOVA models

When we have two factors we automatically have two or more hypotheses of interest. The usual approach is to examine three hypotheses for a two-way ANOVA model. These are (in order of importance)

1. There is no interaction between factor A and factor B

$$H_0 : (\alpha\beta)_{ij} = 0 \text{ for all } i,j$$

Factor label	Definition
BB	beer bottle
BC	beer can
CB	cola bottle
CC	cola can
M	milk
MS	mixed spirits
OJ	orange juice
S	Stolichnaya vodka
W	water

TABLE 5.9: The different beverage and drinking container combinations in Abaz et al.

2. There is no effect due to factor A

$$H_0 : \alpha_i = 0 \text{ for all } i$$

and no effect due to factor B

$$H_0 : \beta_j = 0 \text{ for all } j$$

Note that the last two hypotheses are unordered. That is, there is no reason to test for an effect due to factor A before factor B and vice versa. The α_i's and β_j's are often referred to as the **main effects**. The $(\alpha\beta)_{ij}$ are referred to as the **interaction terms** or **interaction effects**. Therefore, these hypotheses are referred to as "looking at the interactions" and "looking at the main effects."

The important distinction between one-way and two-way ANOVA is the addition of interaction. If the two factors interact, then we *cannot* talk the about the main effects in isolation. That is, if interaction is present, then the response behaves differently for each combination of the factor levels. If there is no interaction present, then we can look at the effect of factor A averaging over the levels of factor B and vice versa.

5.8.2 Example 5.14—DNA left on drinking containers

Abaz et al. [44] were interested in, amongst other things, the effects of person to person variation, type of drink, and type of drink container on the major analytical outcomes of the (DNA) process. As part of their study, they used six volunteers (labeled A to F) and nine different combinations of drinking vessel and beverage, which are listed in Table 5.9. We will treat the beverage/container combinations as different levels of the same factor because the can/bottle variations were not possible for all beverages. We can, at a later point, examine whether the differences between the bottles and cans are significant through the use of contrasts. Each combination of the factor levels has been replicated twice. That is, two measurements have been made on

DNA quantity for each of the 6 × 9 = 54 factor combinations, giving us 108 observations. We use the sum across loci of the heights of the peaks (measured in relative fluorescence units) as proxy for DNA quantity. The justification for this can be found in Abaz et al. [44]. I will justify it here briefly as providing a more stable measurement for the response. This is difficult data to plot because we only have two (or sometimes one) values to plot for each combination of the factors. This lack of data (or difficulty of plotting) is very common in ANOVA. We deal with this by fitting models and examining the behavior of the residuals. Initially, we will fit the *full model* for two-way ANOVA. That is, we will model the data as being affected by both factors and the interaction between them. This model can be written in Wilkinson-Rogers notation as

$$Quantity \sim Person * Container$$

where $Quantity$ is the DNA quantity. This is expanded to

$$Quantity \sim Person + Container + Person \times Container$$

We include the experimental subject ($Person$) as a factor in our model to take account of any differences in the response due to differences in individuals. This is not the *proper* way to do this. We should really **block** on this subject. We will address this matter in the chapter on experimental design. Including the subject as a **treatment factor** (or a **fixed effect**) is often sufficient. Figure 5.36 shows the residual plot from fitting the full two-way ANOVA model. The pred-res plot shows a strong funnel pattern the residuals that indicates a violation of the assumption of equality of variance. This is highlighted in the normal Q-Q plot and the histogram of the residuals where we can see evidence of heavy tails. A log transformation of the data may remove this behavior. I will use a log base 10 transformation instead of the natural logarithm ($\ln x = \log_e x$) because the interpretation is a little more intuitive. That is, it is easier to talk about powers of 10 than it is to talk about powers of Euler's constant $e \approx 2.71828$. Figure 5.37 shows the residuals from the model using the logarithm of $Quantity$ and is much better behaved. The initial examination of the model should be done using the **anova** command. The ANOVA table is displayed in Table 5.10. There are some simple checks that we should perform just to make sure that the data have been treated in the way we want it to be. In particular the degrees of the freedom should be

	Df	Sum Sq	Mean Sq	F value	Pr(> F)
person	5	4.63	0.93	8.26	0.0000
container	8	6.07	0.76	6.78	0.0000
person:container	40	6.83	0.17	1.53	0.0821
Residuals	47	5.26	0.11		

TABLE 5.10: ANOVA table for Abaz et al. data

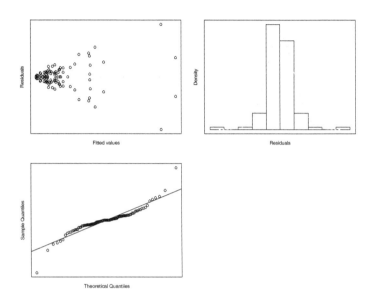

FIGURE 5.36: Diagnostic plots for the two-way ANOVA model on the Abaz et al. data

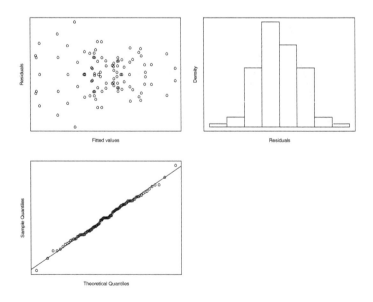

FIGURE 5.37: Diagnostic plots for the two-way ANOVA model on the $\log_{10}$ Abaz et al. data

$I - 1$ and $J - 1$ for each of the main effects, $(I - 1) \times (J - 1)$ for the interactions, and the sum of the degrees of freedom column should be $N_{Tot} - 1$ where N_{Tot} is the total sample size. There are $I = 6$ levels for factor `person` and $J = 9$ levels for factor `container`. We can see that $6 - 1 = 5$, $9 - 1 = 8$, $8 \times 5 = 40$. When we add up the degrees of freedom column we find it only sums to 100, instead of $9 \times 6 \times 2 - 1 = 108 - 1 = 107$. However, recall that there are seven missing values that have not been included in this analysis. We can now look at the P-values for each of the hypotheses. We start at the *bottom* of the ANOVA table with the interaction terms. The P-value is 0.0821, which is weak evidence against the null hypothesis. Given that this adds 40 terms to our model, and complicates the interpretation, we would be better accepting the null hypothesis of no interaction. If we accept that we have no interaction, then we can proceed to interpret each of the P-values for the main effects. We can see that the P-values for the main effects are both very small < 0.0001. Therefore, there is very strong evidence that there is a difference due to the person that the sample was taken from and due to the drinking container/beverage. At this point we could proceed straight to finding where the differences are. However, I recommend that you fit the model you have decided on and use that for interpretation. In theory it should not make any difference. In practice, because of numerical issues, it can often make a slight difference in the estimated effects. It also changes the residual standard error which we use in the calculation of confidence intervals. Therefore, we will fit the **additive model** (the interaction model is sometimes called the **multiplicative model**). At this point we will also fit the model using the `aov` command. The `aov` command allows us to easily get tables of means or effects. It is possible to get them from the regression coefficients but not without a lot more work. The additive model is

$$\log_{10}(Quantity) \sim Person + Container$$

and the ANOVA table from the fit is shown in Table 5.11. The comparisons we can make with an additive model are simpler, because we can *compare the difference in the levels of one factor by averaging over the other*. For example, in this experiment we can examine a hypothesis like "Is there a difference in the DNA quantity for beer bottles and beer cans?" without taking the individual directly into an account. We could not do this if we had chosen an interaction model. We can get tables of means by each factor using the `model.tables` command.

	Df	Sum Sq	Mean Sq	F value	Pr(> F)
person	5	4.63	0.93	6.66	0.0000
container	8	6.07	0.76	5.46	0.0000
Residuals	87	12.10	0.14		

TABLE 5.11: ANOVA table for the additive model for the Abaz et al. data

Tables of means
Grand mean

3.7121

```
person
       A      B      C      D      E      F
    3.544  3.899  3.497  4.068  3.661  3.538
rep 17.000 18.000 16.000 18.000 17.000 15.000

container
      BB     BC     CB     CC     M      MS     OJ     S      W
    3.704  3.383  3.725  4.057  4.019  3.47   3.492  3.555  3.974
rep 9.000 12.000 12.000 12.000 12.000 12.00  10.000 11.000 11.000
```

The output from model.tables is given above. In this instance it returns the grand mean (the average overall observations), the mean for each person averaged over all levels of container, and the mean for each level of container averaged over all people. In addition it tells us how many observations were used in estimating each mean which is important for comparing pairs of means. The formulas for the LSD or the HSD are the same as we saw for one-way ANOVA. We can get an estimate of σ by taking the square root of the residual mean square from the ANOVA table. We will answer the two questions we posed earlier:

1. Is there a difference between beer bottles and beer cans?

2. Is there a difference between cola bottles and cola cans?

The residual mean square from Table 5.11 is 0.1391; therefore, our estimate of σ is $\hat{\sigma} = \sqrt{0.1391} = 0.3729$. To compare μ_{BB} and μ_{BC} we compare (the absolute value of) $\hat{\mu}_{BB} - \hat{\mu}_{BC}$ to

$$LSD = t_{87}(0.025)\hat{\sigma}\sqrt{\frac{1}{n_{BB}} + \frac{1}{n_{BC}}}$$

If the sample difference is greater than LSD, then we say (with 95% confidence) that there is no difference between μ_{BB} and μ_{BC}, or that there is no significant difference at the 0.05 level. For this data set

$$LSD = 1.988 \times 0.3729 \times \sqrt{\frac{1}{9} + \frac{1}{12}}$$
$$= 0.3268$$

The difference between the sample means is $3.704 - 3.383 = 0.321 < 0.3268$. Therefore, the difference is not significant.

Repeating this calculation for the cola bottles and cans we have a difference between the sample means $3.725 - 4.057 = -0.332$. The LSD in this case is

$$LSD = 1.988 \times 0.3729 \times \sqrt{\frac{1}{12} + \frac{1}{12}}$$
$$= 0.3026$$

As $|0.321| > 0.3026$ the difference is significant at the 0.05 level. This difference is calculated on a logarithmic scale. If we give confidence intervals for differences, then it makes sense to express them on their original scales. We can **back transform** our intervals. This changes two things. Firstly, our statements become statements about the median on the original scale rather than the mean. Secondly, the interpretation is **multiplicative** rather than **additive**. We used a $\log_{10}$ transformation in this example, therefore we can back transform by raising the confidence limits to the power 10. E.g., the confidence interval on the $\log_{10}$ scale is

$$\widehat{\mu}_{CB} - \widehat{\mu}_{CC} \pm LSD = -0.332 \pm 0.3026$$
$$= (-0.6347, \ -0.0295)$$

The confidence limits on the original scale are

$$10^{-0.6347} \approx 0.2 \text{ and } 10^{-0.0295} \approx 0.9$$

We interpret this as "The median amount of DNA found on cola bottles is anywhere from 0.2 to 0.9 times lower the median amount of DNA on cola cans." The words "times lower" or "times higher" indicate the multiplicative interpretation.

5.8.3 Tutorial

1. Load the `abaz.df` data frame.

   ```
   > data(abaz.df)
   > names(abaz.df)
   
      [1] "person"         "sample"      "ab.sample"
      [4] "time"           "amylase"     "quant"
      [7] "amp.volume"     "dna.conc"    "gel.profile"
     [10] "failed.profile" "d3"          "vwa"
     [13] "d16"            "d2"          "amelogenin"
     [16] "d8"             "d21"         "d18"
     [19] "d19"            "th01"        "fga"
   ```

2. We need to construct the variable `quantity`. It is formed by summing the peak areas at each locus. The locus values are represented by columns 11 to 21 in the data frame (vWA to FGA). We have a further

complication in that some of the profiles did not develop; hence, the sum of the peak heights will be zero. We want to replace these with missing values NA rather than zeros because otherwise R will not treat them in the way we intend. We will scale this variable by dividing it by 1,000. We do this because the values are quite large and could potentially cause numerical overflow. We will also construct a variable called log.quantity, which is the logarithm base 10 of the unscaled quantity values.

```
> abaz.df = within(abaz.df, {
    quantity = rowSums(abaz.df[, 11:21])
 })
> abaz.df$quantity[abaz.df$quantity == 0] = NA
> abaz.df = within(abaz.df, {
    log.quantity = log10(quantity)
 })
> abaz.df$quantity = abaz.df$quantity/1000
```

3. Information on the different beverage/containers is contained in the variables sample and ab.sample. If we examine the values

```
> xtabs(~person + ab.sample, data = abaz.df)

       ab.sample
person BB BC CB CC M MS OJ R S W
     A  2  2  2  2 2  2  2 1 2 2
     B  2  2  2  2 2  2  2 1 2 2
     C  2  2  2  2 2  2  2 1 2 2
     D  2  2  2  2 2  2  2 1 2 2
     E  2  2  2  2 2  2  2 1 2 2
     F  2  2  2  2 2  2  2 1 2 2
```

you will see that there are two measurements for each beverage/container except the reference R. To simplify the analysis, we will construct a new data frame that omits the reference measurements and keep only those variables of interest.

```
> abaz.df = abaz.df[abaz.df$ab.sample != "R",]
> abaz.df = data.frame(quantity = abaz.df$quantity,
            log.quantity = abaz.df$log.quantity,
            container = factor(
                        as.character(abaz.df$ab.sample)),
            person = abaz.df$person)
```

4. Fit the full two-way ANOVA model with quantity and log.quantity as the response, and person and container as the factors.

The linear model

```
> abaz.fit1 = lm(quantity~person*container,
                 data = abaz.df)
> abaz.fit2 = lm(log.quantity~person*container,
                 data = abaz.df)
```

5. Look at the diagnostic plots for each model.

```
> par(mfrow = c(1, 2), pty = "s")
> plot(abaz.fit1, which = 1:2)
> plot(abaz.fit2, which = 1:2)
```

6. Examine the ANOVA tables for both models.

```
> anova(abaz.fit1)

Analysis of Variance Table

Response: quantity
                 Df Sum Sq Mean Sq F value   Pr(>F)
person            5   2293     459    8.68  6.9e-06 ***
container         8   2148     269    5.08  0.00014 ***
person:container 40   3592      90    1.70  0.04055 *
Residuals        47   2483      53
---
Signif. codes:  0 '***' 0.001 '**' 0.01 '*' 0.05 '.' 0.1 ' ' 1

> anova(abaz.fit2)

Analysis of Variance Table

Response: log.quantity
                 Df Sum Sq Mean Sq F value   Pr(>F)
person            5   4.63   0.926    8.26  1.2e-05 ***
container         8   6.07   0.759    6.78  7.0e-06 ***
person:container 40   6.83   0.171    1.53    0.082 .
Residuals        47   5.26   0.112
---
Signif. codes:  0 '***' 0.001 '**' 0.01 '*' 0.05 '.' 0.1 ' ' 1
```

7. Fit the additive two-way ANOVA model with log.quantity as the response.

```
> abaz.fit3 = lm(log.quantity ~ person + container,
      data = abaz.df)
```

8. Examine the diagnostic plots and the ANOVA table for the additive model.

```
> par(mfrow = c(2, 2))
> plot(abaz.fit3)

> anova(abaz.fit3)
Analysis of Variance Table

Response: log.quantity
          Df Sum Sq Mean Sq F value  Pr(>F)
person     5   4.63   0.926    6.66 2.7e-05 ***
container  8   6.07   0.759    5.46 1.4e-05 ***
Residuals 87  12.10   0.139
---
Signif. codes:  0 '***' 0.001 '**' 0.01 '*' 0.05 '.' 0.1 ' ' 1
```

9. You can use the **anova** command to compare two models where one model has a subset of the terms of the other. Technically, we would say that one model is **nested** within the other.

```
> anova(abaz.fit2, abaz.fit3)
Analysis of Variance Table

Model 1: log.quantity ~ person * container
Model 2: log.quantity ~ person + container
  Res.Df   RSS Df Sum of Sq    F Pr(>F)
1     47  5.26
2     87 12.10 -40     -6.83 1.53  0.082 .
---
Signif. codes:  0 '***' 0.001 '**' 0.01 '*' 0.05 '.' 0.1 ' ' 1
```

10. Get a table of means for each of the main effects (**person** and **container**) using the **model.tables** command. We will need to refit the model with the **aov** command; otherwise, **model.tables** will not work

```
> abaz.fit3 = aov(log.quantity ~ person + container,
     data = abaz.df)
> model.tables(abaz.fit3, "means")

Tables of means
Grand mean

3.7121

 person
         A       B       C       D       E       F
     3.544   3.899   3.497   4.068   3.661   3.538
rep 17.000  18.000  16.000  18.000  17.000  15.000
```

```
       container
             BB     BC     CB     CC      M     MS     OJ      S      W
          3.704  3.383  3.725  4.057  4.019   3.47  3.492  3.555  3.974
      rep 9.000 12.000 12.000 12.000 12.000  12.00 10.000 11.000 11.000
```

11. Extract the mean table and the count for each person.

    ```
    > tbls = model.tables(abaz.fit3, "means")
    > names(tbls)

    [1] "tables" "n"

    > tbls$tables

    $`Grand mean`
    [1] 3.7121

    $person
    person
         A      B      C      D      E      F
    3.5437 3.8994 3.4975 4.0679 3.6611 3.5380

    $container
    container
         BB     BC     CB     CC      M     MS     OJ      S      W
    3.7041 3.3833 3.7246 4.0567 4.0189 3.4699 3.4920 3.5553 3.9743

    > tbls$n

    $person
    person
     A  B  C  D  E  F
    17 18 16 18 17 15

    $container
    container
    BB BC CB CC  M MS OJ  S  W
     9 12 12 12 12 12 10 11 11

    > person.means = tbls$tables$person
    > person.means

    person
         A      B      C      D      E      F
    3.5437 3.8994 3.4975 4.0679 3.6611 3.5380

    > person.counts = tbls$n$person
    > person.counts
    ```

```
person
 A  B  C  D  E  F
17 18 16 18 17 15
```

12. Construct a linear contrast to compare person D to the others. That is, we want to test the hypothesis

$$H_0: \mu_D - \frac{n_A\mu_A + n_B\mu_B + n_C\mu_C + n_E\mu_E + n_F\mu_F}{\sum_{i \neq D} n_i}$$

The contrast coefficients in this case are a little more complicated. The best way to think about the contrast is that we are going to compare the average for person D against the average for all the other people. If the number of observations on each person is not equal, then we cannot just add up the five means and divide by five. Instead, we have to change each mean back into a sum by multiplying by the number of observations for that mean, and then divide by the total number of observations (excluding those on person D). This is equivalent to taking a weighted sum of the means, where the weights are proportional to the group sizes. The test statistic takes the usual form of

$$T_0 = \frac{\widehat{\theta}}{se(\widehat{\theta})}$$

where $\widehat{\theta}$ is the estimate of the contrast. We test this against a Student's t distribution with the degrees of freedom equal to the degrees of freedom for the residual standard error.

Firstly, we will re-order the means and the counts so that person D is the first in the list.

```
> person.means = person.means[c(4, 1, 2, 3, 5, 6)]
> person.counts = person.counts[c(4, 1, 2, 3, 5, 6)]
```

Next, we construct the contrast coefficients and make sure they sum to zero.

```
> person.contrast.c =
    c(1,-person.counts[-1]/sum(person.counts[-1]))
> round(person.contrast.c,2)

     A     B     C     E     F
 1.00 -0.20 -0.22 -0.19 -0.20 -0.18

> sum(person.contrast.c)

[1] 0
```

Now we calculate the estimate

$$\hat{\theta} = \sum_i c_i \hat{\mu}_i$$

```
> person.contrast.est =
    sum(person.contrast.c*person.means)
> person.contrast.est
```

[1] 0.43294

the standard error of the estimate

$$se(\hat{\theta}) = \hat{\sigma}\sqrt{\sum_i \left(\frac{c_i}{n_i}\right)^2}$$

```
> sigma = sqrt(anova(abaz.fit3)["Mean Sq"][3,1])
> person.contrast.se = sigma*sqrt(sum((
            person.contrast.c/person.counts)^2))
> person.contrast.se
```

[1] 0.023025

and finally the test statistic and the P-value

```
> t.stat = person.contrast.est / person.contrast.se
> deg.freedom = anova(abaz.fit3)["Df"][3,1]
> p.value = 2*(1-pt(abs(t.stat),deg.freedom))
> cat(paste(round(t.stat,1),
           deg.freedom,
           round(p.value,4),"\n"))
```

18.8 87 0

We can see from the P-value that there is very strong evidence against the null hypothesis as the test statistic is 18.8 standard deviations above the mean. That is, person D seems to have a much higher average level of DNA than the others in the experiment. The lines

sigma = sqrt(anova(abaz.fit3)["Mean Sq"][3,1])}

and

deg.freedom = anova(abaz.fit3)["Df"][3,1]}

extract the residual standard error and the degrees of freedom of the residual standard error (respectively) from the ANOVA table.

Variables	Technique	$E[y_i|x_i]$
one continuous	Simple linear regression	$\beta_0 + \beta_1 x_i$
one factor	one-way ANOVA	$\mu + \alpha_i$
two factors	two-way ANOVA	$\mu + \alpha_i + \beta_j + (\alpha\beta)_{ij}$
two or more variables: mixture of continuous and factors	multiple regression	$\beta_0 + \beta_1 x_{1i} + \cdots + \beta_p x_{pi}$

TABLE 5.12: The mean for a linear model by technique

5.9 Unifying the linear model

There is a useful unifying concept for linear models. Every linear model can be written as

$$y_i = E[y_i|x_i] + \epsilon_i, \ \epsilon_i \sim N(0, \sigma^2) \tag{5.16}$$

where the form of $E[y_i|x_i]$ changes according the nature of the explanatory variables, or the particular modeling framework in which we wish to interpret our data. The key fact is that the form of $E[y_i|x_i]$ is always linear for the linear model. These specific forms are listed in Table 5.12. This unifying concept is useful for two reasons. Firstly, it allows us to see that every linear model has the same form and hence the same tools can be used in every analysis. Secondly, it makes it very clear that we are modeling the mean (or expected value) of y_i given x_i as a linear function of the explanatory variables. This will prove to be a useful idea in the next chapter.

5.9.1 The ANOVA identity

Every model, linear or not, that we encounter in this book can be thought of as

$$DATA = SIGNAL + NOISE$$

One way of thinking about this in a more statistical sense is to think about the attribution of variation to sources. That is, there is a certain amount of inherent variability in the data. When we model the data, we attempt to "explain," "assign," or "attribute" variability with explanatory variables. If the variables explain the data particularly well, then they will "soak up" a large proportion of the variability. This concept is encapsulated for linear models in what is called the **ANOVA identity**. The ANOVA identity usually expressed in terms of one-way analysis of variance, but it works for any linear model. We write

$$Total \ SS = Model \ SS + Residual \ SS$$

where

- *Total SS* is the **total sum of squares**. It is an unscaled measure of the total variability present in the data

- *Model SS* is the **model sum of squares**. It is measure of the variability in the data that is explained by the model

- and *Residual SS* is the **residual sum of squares** or sometimes the **error sum of squares**. It represents the variability in the data not explained by the model

It is not hard to see that we can use this decomposition to derive R^2.

$$\begin{aligned} R^2 &= \frac{MSS}{TSS} \\ &= \frac{TSS - RSS}{TSS} \\ &= 1 - \frac{RSS}{TSS} \end{aligned}$$

The model sum of squares is not usually presented on a single line in the ANOVA table for a linear model. However, it is easily calculated by summing of all the terms in the sum of squares column above `residual`. Similarly, R does not give the total sum of squares, but it can be easily calculated by adding up all the entries in the sum of squares column.

When there is more than one variable present in the model, the model sum of squares is broken down for each variable in the model. Note that this for each variable, not for each regression parameter, so a one-way ANOVA will only have one entry for the factor rather than one for each variable. This decomposition can sometimes help us assess the value of adding or removing terms from a model. This is best illustrated with an example.

Recall the data from Gustafson [1] in section 5.4.3. The ANOVA table from fitted the full model is given in Table 5.13. The residual sum of squares

	Df	Sum Sq	Mean Sq	F value	Pr(> F)
A	1	4325.69	4325.69	82.03	0.0000
P	1	1069.41	1069.41	20.28	0.0001
C	1	2364.10	2364.10	44.83	0.0000
R	1	733.93	733.93	13.92	0.0007
T	1	1034.60	1034.60	19.62	0.0001
S	1	191.64	191.64	3.63	0.0651
Residuals	34	1792.86	52.73		

TABLE 5.13: ANOVA table for the full multiple regression model on Gustafson's tooth data

is 1792.9. The total sum of squares is 11512.2 and is obtained by adding up the "Sum Sq" column in Table 5.13. We can get the model sum of squares by

	Sum Sq	%	Cum. %
A	4325.69	37.6	37.6
P	1069.41	9.3	46.9
C	2364.10	20.5	67.4
R	733.93	6.4	73.8
T	1034.60	9.0	82.8
S	191.64	1.7	84.4
Residuals	1792.86	15.6	100.0

TABLE 5.14: Model terms as a percentage of (cumulative) variance explained

subtraction, i.e., $9719.4 = 11512.2 - 1792.9$. We can also, however, express the sum of squares for each model term as a percentage of the total sum of squares. This is shown in Table 5.14. We can get a sense of how much each term is *contributing* to the fit. So, for example, the addition of the S term only adds 1.7% to the percentage of variation explained. Therefore, we might feel justified in dropping it from the model, which we did in the original analysis. If you are re-reading the original analysis, you might recall that we also dropped the variable A. This type of analysis is *sequential*. That is, the order in which variables are added affects the sum of squares. So you do need to be careful in your interpretation. If you are planning to remove a variable on the basis of this type of analysis, make sure it is the last variable to enter the model. Having said that, this type of analysis is particularly useful when interaction terms are involved. The reason it is useful is that interaction terms make model interpretation more difficult. Significant interaction terms can also be made significant by just one specific interaction. For example, we might have a model with two factors, one with three levels and the other with four levels. That means that in the interaction model, there will be $4 \times 3 = 12$ interaction means. If *just one* of those means is significantly different from the others, then the F-test for interaction will be significant. Looking at the percentage of variation explained by the interaction terms gives us a way to judge the usefulness of adding in this extra model complexity. We can generally have faith in this situation because interaction terms are almost always added to the model last.

Chapter 6

Modeling count and proportion data

We define the art of conjecture, or stochastic art, as the art of evaluating as exactly as possible the probabilities of things, so that in our judgments and actions we can always base ourselves on what has been found to be the best, the most appropriate, the most certain, the best advised; this is the only object of the wisdom of the philosopher and the prudence of the statesman. – Jacob Bernoulli, Mathematician.

6.1 Who should read this?

This chapter provides an introduction to models that allow us to deal with count data and proportion data. If the response in your experiment or study is a count of something, then this section will provide some models for this kind of data. For example, we might count the number of glass fragments on the ground after breaking a series of windows. We will discuss the use of Poisson and negative binomial regression models for this type of data. Alternatively, if your data have been cross-classified into two (or more categories) and you wish to know whether the odds of being in one category or another is affected by one or more variables, then this section will provide models for this kind of data. For example, we might wish to know whether the odds on suicide are affected by gender and genotype.

6.2 How to read this chapter

This chapter provides a brief, non-technical introduction to **generalized linear models** (GLMs). In particular we will discuss Poisson, negative binomial and logistic regression. These are Poisson, negative binomial and binomial GLMs respectively.

If the response in your experiment takes values of $0, 1, 2, \ldots$ then you should read the sections on Poisson and negative binomial regression. Note that these models can deal with data where observing a value of zero is not technically possible, and there may be a finite (but large) upper limit. The key point is that we are interested in modeling the counts.

If the response in your experiment is

- binary, i.e., only has two possible outcomes

- categorical, i.e., can fall into one of k categories

- a proportion of successes, i.e., a value between 0 and 1

then you should read the section on logistic regression. The key point is that we are interested in modeling the properties that govern membership of two (or more in the multinomial case) categories.

The explanatory variables may be a mixture of quantitative and qualitative variables in both cases.

6.3 Introduction to GLMs

The generalized linear model is, as the name suggests, a generalized framework for linear models. Confusingly, generalized linear models are distinct from **general linear models**. The concept is relatively easy to understand. The unifying concept for the linear model was that *all* linear models take the form

$$y_i = E[y_i|x_i] + \epsilon_i, \;\; \epsilon_i \sim N(0, \sigma^2)$$

where the form of $E[y_i|x_i]$ will depend on the specific model (simple/multiple regression, one-way/two-way ANOVA) we are considering. Another way of thinking of these models is that we believe the response comes from a specific distribution and we are modeling the expected value of Y given one or more X. For the linear model, we assume/believe that the response is normally distributed with a mean that is a linear function of the explanatory variables. That is,

$$y_i \sim N(\mu_{x_i}, \sigma^2)$$

where

$$\mu_{x_i} = E[y_i|x_i] = \beta_0 + \beta_1 x_{1i} + \beta_2 x_{2i} + \cdots + \beta_p x_{pi}$$

A natural question is, "What do we do if we do not think our data are normal?" The answer is that we use a generalized linear model.

6.4 Poisson regression or Poisson GLMs

The distribution of counts of items or objects may be modeled by the normal distribution, but given the discrete nature of the data it may be more appropriate to use the Poisson distribution. The Poisson distribution models random variables with outcomes $0, 1, 2, \ldots$. The distribution is governed by its mean λ. The idea with Poisson regression is that we believe or assume that the response follows a Poisson distribution where the mean is different for different values of the explanatory variable. That is,

$$y_i \sim Poisson(\lambda_{x_i})$$

where

$$\log(\lambda_{x_i}) = \log(E[y_i|x_i]) = \beta_0 + \beta_1 x_{1i} + \beta_2 x_{2i} + \cdots + \beta_p x_{pi}$$

You can see that this is very similar to a linear model. The major difference is that we assume the response is Poisson distributed instead of normally distributed. Notice also that in this situation we assume that the logarithm of the mean is a linear function, rather than just the mean itself. The log function is known as the **canonical link function** for the Poisson distribution. In general, a **link function** provides a *link* between the mean and the linear predictor (the linear function). The choice of link functions is somewhat arbitrary, but usually there are common choices for each distribution, called the canonical link function. The log link is the canonical link for the Poisson distribution. The identity link (which means we do nothing) is the canonical link for the normal distribution and so on.

6.4.1 Example 6.1—Glass fragments on the ground

Wong [45] was interested in the modeling the number of glass fragments on the ground that were recovered after a window pane was shot with a handgun. In particular, he explored the hypothesis that the (mean) number of fragments was altered by the velocity, hardness, and shape (profile) of the projectile – the projectile being the lead part of a bullet. Wong, Buckleton, and Curran (*pers. comm.*) designed an experiment to investigate this hypothesis. The projectile velocity was controlled by altering the amount of gunpowder added to each bullet. The hardness (as measured on the Rockwell scale of hardness) of each projectile was altered by changing the amount of antimony (Sb) added to the projectile lead during casting. The profile of the projectile was changed by using a round-nose (RN) or wad-cutter (WC) mold. A full factorial design was used to allocate combinations of the factors (velocity, hardness, and profile) to the experimental units (shots). There were four velocity levels, three hardness levels, and two profile levels. Therefore, to make sure every combination of the factors is repeated once requires $4 \times 3 \times 2 = 24$ experiments. In order to understand the variability in the process, we need at least two or more replicate firings for each combination of the factors in a fully factorial design. Wong chose to repeat each combination of factors three times, which would result in $3 \times 24 = 72$ observations. Of course, in any experiment, there are some cases where things just do not work. Wong found that there was a problem with the projectiles either deforming or disintegrating under certain conditions. In addition some projectiles were unstable in flight and missed the target altogether. As a result, only 54 of the planned 72 experiments were carried out. The velocity of the projectiles is only controllable up to a point. For that reason, the velocity of each projectile was measured using a chronograph. This means in our data set we have two factors (P = profile and H = hardness) and one continuous variable (V = velocity). In this example, we will not consider the interactions between the variables, although in general it would make sense to do so.

We fit the model using the `glm` command.

```
> data(wong.df)
> wong.fit = glm(Count ~ V + H + P, data = wong.df,
    family = poisson)
```

You can see from the syntax that the model specification is exactly the same as we used for `lm`. There is an additional parameter `family` which lets us tell R which family of distributions we are using to model the data. In this case we are using the Poisson distribution. The specification of the family also tells R to use the canonical link function for that family. It is possible to specify another link function or even your own link function, but that is beyond the scope of this book.

It is possible to produce diagnostic plots for a GLM which are similar to those produced for a linear model. However, they can be difficult to in interpret and/or misleading and we will avoid them here.

The significance or otherwise of variables in a GLM is assessed by examination of the **analysis of deviance table**. We will explain the concept of deviance shortly, but an interim way of thinking about it is to regard it as being similar to the residual sum of squares for a linear model. We start with a base level of variability in the response. This base level is described by the total sum of squares which is an unscaled measure of the variability of the response around the mean. That is, the total sum of squares for a linear model is

$$Total\ SS = \sum_i (y_i - \bar{y})^2$$

This measures the variability of the data in the model

$$E[Y_i] = \mu$$

where $\hat{\mu} = \bar{y}$. As we add terms to our model, some of this variability is assigned to the model sum of squares and what is left over is called the residual sum of squares. Therefore, the addition of terms to the model increases the model sum of squares and reduces the residual sum of squares. We can use the F distribution to test the hypothesis that the reduction in the residual sum of squares caused by the addition of the explanatory variable to the model is too large to have been due to random variability alone. That is, we can use the changes in the residual sum of squares to (sequentially) test the significance of explanatory variables. The same is true of deviance. The null model is the model where the explanatory variables have no effect on the response. Associated with the null model is a deviance. As terms are added to the model, this deviance will decrease. The changes in deviance can be compared to a χ^2 distribution. If the change is "too large," then we regard this variable as being potentially important in explaining the response. It is important to know that this is a sequential process, and so the order in which variables are added can affect the significance of subsequent variables. Variable selection procedures for GLMs are not as well developed as they are for linear models. The only practical advice I can give is that if you think a variable is significant, then refit the model again with the variable in question entering the model last. If the variable remains significant in the analysis of deviance table, then it is probably important.

The analysis of deviance table for Wong's data can be obtained (rather strangely or comfortingly depending on your point of view) with the anova command. The resulting analysis of deviance table is shown in Table 6.1. We can see that all three variables are very important in predicting the response. Table 6.2 shows how the analysis of deviance table changes if we permute the order in which the variables enter the model, to $Count \sim H + P + V$, i.e., H and P enter the model before V this time. We can see that V still has by far the

	Df	Deviance	Resid. Df	Resid. Dev.	$\Pr(>\chi^2)$
NULL			53	261781.59	
V	1	205238.18	52	56543.41	0.0000
H	2	23376.69	50	33166.71	0.0000
P	1	3932.95	49	29233.76	0.0000

TABLE 6.1: Analysis of deviance table for Poisson regression on Wong data

	Df	Deviance	Resid. Df	Resid. Dev.	$\Pr(>\chi^2)$
NULL			53	261781.59	
H	2	12459.10	51	249322.49	0.0000
P	1	77778.20	50	171544.29	0.0000
V	1	142310.52	49	29233.76	0.0000

TABLE 6.2: Analysis of deviance table for Poisson regression on Wong data with the variables re-ordered

largest effect on the deviance. Variables H and P still have a large effect, but you will notice that the relative sizes in the change in deviance have reversed. This in itself is unimportant except that it shows the sequential nature of the process. We can examine how well our model explains our data with a number of numerical measures. However, it is worth initially just plotting the fitted counts versus the observed counts. This will give us a nice graphical summary of the fit. Given the nature of the data (extreme right skew) it makes sense to plot on a logarithmic scale rather than the raw count scale. Figure 6.1 shows that there appears to be a strong concordance between the observed values and the fitted values (the correlation is 0.97). However, the pred-res plot on the original scale tells a slightly different story. A feature of the Poisson distribution is that the variance is equal to the mean. That is, if $X \sim Poisson(\lambda)$ then the mean of X is $E[X] = \lambda$ and the variance of X is $Var[X] = \lambda$. Therefore, in a Poisson regression, we expect the variation about the line (the mean) to increase at the same rate as the line. If the variation increases faster than the mean, then we say that the data are **over-dispersed**. That is, over-dispersion means that we observe more variation in the data than we expect from the Poisson model. Similarly, if the variation does not increase as fast as the mean, then we say that the data are **under-dispersed**. We can examine this property in Poisson regression by plotting the fitted values against the variance. Following Faraway [46], the variance for a particular value of x is hard to estimate, but the squared difference between the observed and fitted value $(y_i - \widehat{y}_i)^2$ serves as a crude proxy. This plot is shown in Figure

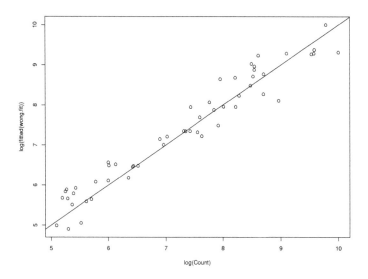

FIGURE 6.1: Plot of fitted versus observed values for the Wong data

6.3. The line indicates the Poisson assumption of mean/variance equality. We

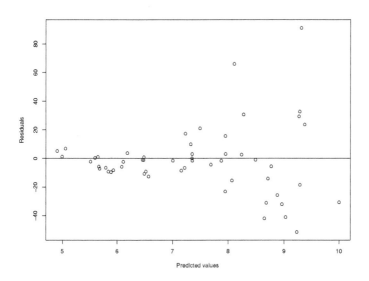

FIGURE 6.2: Pred-res plot for the Poisson regression on Wong's data

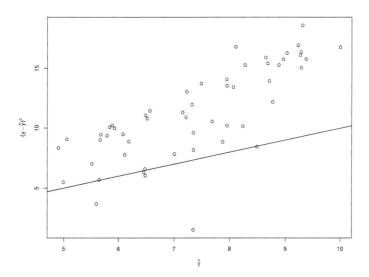

FIGURE 6.3: The relationship between the mean and the variance for the Poisson model of the Wong data

can see from Figure 6.3 that there appears to be a linear relationship between the mean and the variance, but the variance is increasing at a faster rate than the mean, which suggests that the data are over-dispersed. The consequences of this are that although the estimates of the coefficients will be **consistent**, the standard errors will be wrong. Consistency of the estimates meaning that for a big enough sample size they will be close to the true values. Therefore, we cannot say which variables will be important. We can look at the residuals to see if an outlier is the cause of this over-dispersion. We can use a **half-normal plot** for this purpose. The half-normal plot plots absolute values of the data (or residuals) against the positive quantiles of the normal distribution. It is similar to the normal Q-Q plot, but lets us focus on the magnitude of the values instead of the sign. That is, we are interested in finding values that are a long way from the mean—we do not care if they are above or below the mean. There are some large residuals in Figure 6.4. We can assess the impact on the fit if we remove the points which have residuals larger than 40 in absolute value. The choice of 40 is arbitrary, but it does look like there is a bit of a jump in the residuals above that point. We need something to judge whether the removal of these points helps or not. We can do this by using the residual deviance. The residual deviance is an entire column in the analysis of deviance table. When I refer to the residual deviance with no further qualification I mean the residual deviance after all the variables in the model have been fitted. This is the residual deviance on the last line of the analysis of deviance table. Alternatively, if we are using the summary

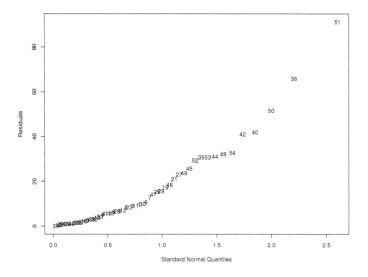

FIGURE 6.4: Half-normal plot for the residual from the Poisson model

Modeling count and proportion data

command, it will appear at the bottom of the output as `Residual deviance` or it can be obtained with the `deviance` function. We compare the residual deviance to the 95% critical point from the χ^2 distribution with degrees of freedom equal to the residual deviance degrees of freedom (ν). If the residual deviance is less than $\chi_\nu(0.95)$, then for reasons we will not go into yet, the Poisson model is deemed to be adequate. We can see from first model that the residual deviance is 29230 on 49 degrees of freedom. The 95% critical value for the χ^2 distribution on $\nu = 49$ degrees of freedom is 66.34, therefore the model is a very poor fit. In fact we can use this criterion as a measure of model fit. If D_{res} is the residual deviance and ν are the residual degrees of freedom, then

$$1 - \Pr(X^2 < D_{res})$$

where X^2 is a χ^2 random variable with ν degrees of freedom can be used in an analogous way to R^2. If you only have the analysis of deviance table and R is not easily to hand, a quick eyeball check is that the residual deviance should be close to the residual degrees of freedom. Lindsey [47] suggests no more than twice the degrees of freedom. This works because the mean of a χ^2 random variable with $\nu > 2$ is ν and its variance is 2ν. It can be shown that $\nu + 2\sqrt{2\nu}$, which is the upper bound of a 2 standard deviation interval, is less than 2ν for $\nu > 8$. We can see from Table 6.3 that the residual deviance we get from refitting the model with all of the potential outliers removed is 9911 on 44 degrees of freedom. We can tell with our eyeball check that the model is still a very poor fit.

6.5 The negative binomial GLM

We have established that the Wong data are over-dispersed for the Poisson model. However, the linear relationship in Figure 6.3 showed us that although the mean-variance equality assumption for the Poisson was violated, the link function was correctly specified. What we would like to do is modify the Poisson model so that the mean is still $E[Y_i|X_i] = \lambda_i$, but the variance is $Var[Y_i|X_i] = \phi \lambda_i$. ϕ is called the **dispersion parameter**. This can be achieved by letting the mean of the Poisson λ be a gamma distributed random

	Df	Deviance	Resid. Df	Resid. Dev.
NULL			48	208297.89
V	1	169663.64	47	38634.26
H	2	24917.82	45	13716.44
P	1	3804.94	44	9911.50

TABLE 6.3: Analysis of deviance table for Wong data with outliers removed

variable with mean μ and variance μ/ϕ. This in itself will not be important to most readers, but the net result is that the response now follows a **negative binomial** distribution, and hence motivates the negative binomial generalized linear model for over- or under-dispersed count data.

6.5.1 Example 6.2—Over–dispersed data

We will fit a negative binomial model to the Wong data. To highlight the usefulness of being able to deal with over-dispersion, we will firstly fit a Poisson model to the Wong data, as we did before. However, this time we will fit the saturated model. That is, we will fit the model

$$Count_i \sim Poisson(\lambda_i)$$
$$\log(\lambda_i) = V \times H \times P$$

Remember that $V \times H \times P$ expands into a model with all the main effects for V, H, and P, plus all the two-way interactions between those variables, plus the three-way interaction. The analysis of deviance table for this model is shown in Table 6.4. Immediately obvious from Table 6.4 is the fact that, if we believe the χ^2 tests, the three-way interaction term V:P:H is significant. This means that if we accept this model we have six different intercepts, one for each combination of hardness and profile. This is clearly a complicated model to explain. We can simplify the model, as Wong [45] did, on a "percentage of deviance" argument. In such an argument, we regard the null deviance as being similar to the total sum of squares. The addition of all the interactions to the model reduces the residual deviance from 29234 to 25477, a change of 3757. This figure, expressed as a percentage of the null deviance is 1.4%. This absolutely pales into insignificance when we compare it to the percentage for velocity (78.4%), and hence we adjudge the extra model complexity to "not be worth the trouble."

An alternative approach is to adjust the significance tests for the over-dispersion. This is achieved by the using the **quasi-Poisson** family of dis-

	Df	Deviance	Resid. Df	Resid. Dev.	$\Pr(> \chi^2)$
NULL			53	261781.59	
V	1	205238.18	52	56543.41	0.0000
H	2	23376.69	50	33166.71	0.0000
P	1	3932.95	49	29233.76	0.0000
V:H	2	3059.64	47	26174.12	0.0000
V:P	1	6.04	46	26168.08	0.0140
H:P	2	677.34	44	25490.74	0.0000
V:H:P	2	13.54	42	25477.20	0.0011

TABLE 6.4: A fully saturated Poisson model for the Wong data

tributions. I will not go into the details of this distribution. However, if we estimate the dispersion parameter ϕ by

$$\widehat{\phi} = \frac{\sum_i (\widehat{y}_i - y_i)^2 / \widehat{y}_i}{\nu}$$

where ν is the residual degrees of freedom, then we can scale the residual deviance column in the analysis of deviance table to construct **pseudo-F statistics**, which we can then use in conjunction with the F distribution to perform significance tests. This sounds complicated, but it can be achieved in one step by using the family = quasipoisson option in the GLM call. The analysis of deviance table from such a fit is shown in Table 6.5. We can see from Table 6.5 that performing such a correction allows us to justify ignoring the interaction terms. However, this correction makes no change to the coefficients of the model, and hence we have the same problems with the residuals. Having said that, the quasi-Poisson model may be preferable to the negative binomial because it assumes a linear relationship between the mean and the variance. The negative binomial on the other hand assumes a quadratic relationship.

The negative binomial explicitly allows for the over-dispersion rather than simply trying to correct for it. To do this, it requires an additional parameter that must be estimated. Fortunately R can do this for us. Under the negative binomial model, we assume

$$Y_i \sim Neg.\ Bin(\mu_i, \theta)$$
$$\log(\mu_i) = \beta_0 + \beta_1 x_{1i} + \cdots + \beta_p x_{pi}$$

The log link is not the canonical link for the negative binomial, but it is the most commonly used. The canonical link is hard to interpret and hence not commonly used. R allows the inverse and square root links to be used as well. To fit a negative binomial GLM, we need to load the MASS library so that we can use the glm.nb command. MASS stands for "Modern Applied Statistics

	Df	Deviance	Resid. Df	Resid. Dev.	F value	Pr(> F)
NULL			53	261781.59		
V	1	205238.18	52	56543.41	320.47	0.0000
H	2	23376.69	50	33166.71	18.25	0.0000
P	1	3932.95	49	29233.76	6.14	0.0173
V:H	2	3059.64	47	26174.12	2.39	0.1041
V:P	1	6.04	46	26168.08	0.01	0.9231
H:P	2	677.34	44	25490.74	0.53	0.5932
V:H:P	2	13.54	42	25477.20	0.01	0.9895

TABLE 6.5: Analysis of deviance table for the Wong data using a quasi-Poisson model

with S" and is written for and extensively used in the book of the same name by Venables and Ripley [48]. The MASS library is included in the standard installation of R, so we do not need to download and install it to use it. We use `glm.nb` because it provides simultaneous estimation of θ thereby obviating us from the task.

```
Analysis of Deviance Table

Model: Negative Binomial(10.877), link: log

Response: Count

Terms added sequentially (first to last)


      Df Deviance Resid. Df Resid. Dev P(>|Chi|)
NULL                   53       1039
V      1      895     52        145   < 2e-16 ***
H      2       60     50         84   8.8e-14 ***
P      1       21     49         64   5.6e-06 ***
V:H    2        4     47         60    0.14
V:P    1        1     46         59    0.45
H:P    2        1     44         58    0.58
V:H:P  2        4     42         54    0.14
---
Signif. codes:  0 '***' 0.001 '**' 0.01 '*' 0.05 '.' 0.1 ' ' 1
```

The analysis of deviance table for the negative binomial model is given above. We can see that once we have taken care of the over-dispersion, by fitting a model that explicitly allows the variance to increase faster than the mean, that none of the interaction terms are significant. This means that we should explore fitting the additive model to the data. We can use the `anova` command to compare the two models as we did with linear models. The additive model is $Count \sim V + H + P$. We fit it in the usual way.

```
> wong.nb1 = glm.nb(Count ~ V + H + P, data = wong.df)
> anova(wong.nb, wong.nb1)

Likelihood ratio tests of Negative Binomial Models

Response: Count
      Model     theta Resid. df   2 x log-lik.   Test
1 V + H + P    9.3094        49        -820.33
2 V * H * P   10.8772        42        -811.44 1 vs 2
     df LR stat.  Pr(Chi)
1
2     7   8.8937  0.26038
```

We can see from the output above that the *P*-value (0.26) is not even close to significance; therefore ,the reduced (additive) model is adequate.

6.5.2 Example 6.3—Thoracic injuries in car crashes

Kent and Patrie [49] where interested in the development of thoracic injury risk functions with respect to a variety of variables such as age, gender and the loading condition in car crashes. They collected data sets from crash experiments that met various criteria. The resulting data set has 93 observations. The four loading conditions are grouped as

1. blunt hub (41 tests)
2. seatbelt (26 tests)
3. distributed (12 tests)
4. combined belt and bag (14) tests

Full descriptions of each of these groupings are given in the original paper. Each experiment recorded the total number of rib fractures, the loading, the age and gender, and the maximum level of chest deflection (Cmax). Given that the number of rib fractures is a count variable, we might reasonably assume that it could be modeled with a Poisson distribution. The variables gender and load.cond are factor variables with 2 and 4 levels, respectively. The variables cmax and age are continuous. To fit the fully saturated model therefore, with every combination of variables will require 32 parameters. We have 93 observations so we have sufficient data to fit the model, even though it may not be very sensible. The analysis of deviance table is shown in Table 6.6. From the table we can see that three of the four *main effects* are significant as well as a number of the higher-order interactions. However, the deviance after all the model terms have been fitted is 200.16. This is more than three times larger than the 61 degrees of freedom. Therefore, we might believe that the significance of the higher-order interaction terms has more to do with over-dispersion than these interactions actually being important in explaining the response. This belief is confirmed when we adjust for the over-dispersion by using the quasi-Poisson family. We can see from Table 6.7 that only age, gender, and cmax appear to be having any effect. cmax absorbed the largest portion of the variation suggesting that it might be the most important predictor.

6.5.3 Example 6.4—Over-dispersion in car crash data

The Kent and Patrie [49] rib fracture data are over-dispersed under the Poisson model. Therefore, we might consider modeling the data with a negative binomial GLM to take account of the over dispersion. We fit the fully saturated model again. Table 6.8 shows the analysis of deviance table for this model. Interestingly, there appear to be a few significant two- and three-way interactions. In this case I will argue that the addition of 25 extra model parameters to get marginal gains in performance is probably not worth the extra complication.

	Resid. Df	Resid. Dev.	$\Pr(>\chi^2)$
NULL	92	587.71	
age	91	555.15	0.0000
gender	90	514.24	0.0000
cmax	89	305.99	0.0000
load.cond	86	298.39	0.0552
age:gender	85	296.44	0.1622
age:cmax	84	290.23	0.0127
gender:cmax	83	290.01	0.6397
age:load.cond	80	273.92	0.0011
gender:load.cond	77	253.34	0.0001
cmax:load.cond	74	234.37	0.0003
age:gender:cmax	73	231.49	0.0897
age:gender:load.cond	70	227.97	0.3191
age:cmax:load.cond	67	222.24	0.1253
gender:cmax:load.cond	64	208.00	0.0026
age:gender:cmax:load.cond	61	200.16	0.0494

TABLE 6.6: Analysis of deviance table for the fully saturated Poisson GLM fitted to the Kent and Patrie data

	Df	Deviance	F	$\Pr(>F)$
NULL				
age	1	32.56	11.06	0.0015
gender	1	40.91	13.89	0.0004
cmax	1	208.25	70.70	0.0000
load.cond	3	7.59	0.86	0.4671
age:gender	1	1.95	0.66	0.4185
age:cmax	1	6.21	2.11	0.1517
gender:cmax	1	0.22	0.07	0.7860
age:load.cond	3	16.09	1.82	0.1528
gender:load.cond	3	20.58	2.33	0.0832
cmax:load.cond	3	18.98	2.15	0.1035
age:gender:cmax	1	2.88	0.98	0.3266
age:gender:load.cond	3	3.51	0.40	0.7552
age:cmax:load.cond	3	5.73	0.65	0.5867
gender:cmax:load.cond	3	14.24	1.61	0.1959
age:gender:cmax:load.cond	3	7.84	0.89	0.4528

TABLE 6.7: Analysis of deviance table for the fully saturated quasi-Poisson GLM fitted to the Kent and Patrie data

6.5.4 Tutorial

1. Load the Wong data.

   ```
   > data(wong.df)
   > names(wong.df)

   [1] "Expt"   "V"   "H"   "P"   "Count"
   ```

2. Add two additional variables to the data frame—the logarithm of the count, and the square root of the count. In addition remove the variable Expt.

   ```
   > wong.df = wong.df[, -1]
   > wong.df = within(wong.df, {
         log.count = log(Count)
         sqrt.count = sqrt(Count)
     })
   > wong.df = wong.df[, c(4:6, 1:3)]
   ```

   The last line of code reorders the columns of the data frame so that the response variables (Count) and the two transformed response variables are in the first columns. We do this so that the response variables are the lead entries in the pairs plot.

3. Load the s20x library and produce a pairs plot of the variables

	Resid. Df	Resid. Dev.	$\Pr(> \chi^2)$
NULL	92	307.01	
age	91	289.56	0.0000
gender	90	265.78	0.0000
cmax	89	175.47	0.0000
load.cond	86	171.59	0.2747
age:gender	85	170.70	0.3455
age:cmax	84	168.23	0.1162
gender:cmax	83	167.21	0.3124
age:load.cond	80	160.97	0.1006
gender:load.cond	77	150.76	0.0169
cmax:load.cond	74	136.40	0.0024
age:gender:cmax	73	133.64	0.0967
age:gender:load.cond	70	132.19	0.6939
age:cmax:load.cond	67	128.07	0.2494
gender:cmax:load.cond	64	117.21	0.0125
age:gender:cmax:load.cond	61	113.42	0.2845

TABLE 6.8: Analysis of deviance table for the fully saturated negative binomial GLM fitted to the Kent and Patrie data

```
> library(s20x)
> pairs20x(wong.df)
```

We can see from Figure 6.5 that the log response is quite linear with respect to velocity, and that there is a quadratic relationship with the square root of the response. A traditional approach to count data is to model the square root of the counts with as a linear function of the explanatory variables. We will contrast this model, and a linear model with a log response, to the others that we have explored in this chapter.

4. Fit linear models to the square root of the count and the log of count. Include a quadratic term for velocity in the square root model.

```
> wong.lm.sqrt = lm(sqrt.count ~ V + I(V^2) +
    H + P, data = wong.df)
> wong.lm.log = lm(log.count ~ V + P + H, data = wong.df)
```

5. Examine the diagnostic plots for each of these model and the ANOVA and summary tables.

```
> plot(wong.lm.sqrt)
> plot(wong.lm.log)
> anova(wong.lm.sqrt)
> anova(wong.lm.log)
> summary(wong.lm.sqrt)
> summary(wong.lm.log)
```

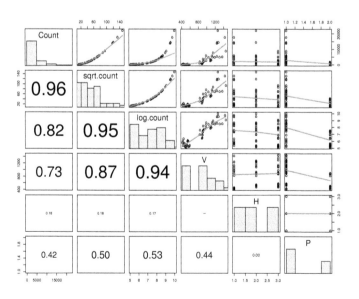

FIGURE 6.5: A pairs plot for the Wong data

Modeling count and proportion data

We can see from the diagnostic plots that the square root transformation does not work at all. There is a strong non-linearity and non-constant variance in the pred-res plot and evidence of long tails in the normal Q-Q plot. These problems are lessened considerably in the log response model, and herein lies an important lesson—you can often achieve good results with the "wrong model."

6. Fit the saturated and additive Poisson regression models

```
> wong.glm.pois = glm(Count ~ V * H * P, data = wong.df,
    family = poisson)
> wong.glm.pois1 = glm(Count ~ V + H + P, data = wong.df,
    family = poisson)
```

7. Get the analysis of deviance tables for both models. To perform the χ^2-tests of significance, we need to specify `test = "Chisq"`. Note that R is very particular about the capitalization here and that `test = "chisq"` will give you an error.

```
> anova(wong.glm.pois, test = "Chisq")
> anova(wong.glm.pois1, test = "Chisq")
> anova(wong.glm.pois, wong.glm.pois1, test = "Chisq")
```

The last line of code here compares the two models and tests the hypothesis that none of the interaction terms is significant. If the interactions are significant in the full model, then they will be significant in this test.

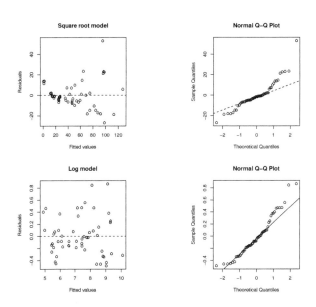

FIGURE 6.6: Diagnostic plots for linear models fitted to the Wong data

8. It is possible to use the `plot` command on the fitted `glm` objects in the same way that we do for the `lm`. Remember that you should treat these plots with some degree of caution.

```
> plot(wong.glm.pois)
> plot(wong.glm.pois1)
```

9. Another useful diagnostic plot is the half-normal plot. To construct this plot, we plot the sorted absolute value of the residuals against the quantiles from the standard normal distribution that are greater than 0.5. There are a variety of different ways of selecting the α values for the quantiles. Faraway [50] uses

$$\alpha_i = \frac{n+i}{2n+1}$$

where n is the number of observations. An alternative is to use the R function `ppoints` with argument $2n+1$ and select only those points that are greater than 0.5. Either method is satisfactory. Rather than repeat the code for each model I have wrapped the code for the half-normal plot in a function so that we can use it again. The function `halfnorm` takes a fitted `lm` or `glm` object as its argument.

```
> halfnorm = function(fit) {
    n = length(residuals(fit))
    p = ppoints(2 * n + 1)
    alpha = p[p > 0.5]
    absRes = abs(residuals(fit))
    resLabs = as.character(1:n)
    o = order(absRes)
    absRes = absRes[o]
    resLabs = resLabs[o]
    plot(qnorm(alpha), absRes,
        xlab = "Standard normal quantiles",
        ylab = "abs(Res)")
    big = which(alpha > 0.95)
    if (length(big) > 0)
        text(qnorm(alpha[big]), absRes[big],
            resLabs[big], cex = 0.7, adj = c(1,
                0))
}
> halfnorm(wong.glm.pois)
> halfnorm(wong.glm.pois1)
```

This function is also in the `dafs` package. The half-normal plots for each model are shown in Figure 6.7. The code above labels points where $\alpha > 0.95$ for no other reason than this is a convenient way to label the highest 5% of points. However, it does let us identify the extreme observations. The highest points are points 38 and 51. As we know these models are unsatisfactory because of over-dispersion we will not go through the steps to re-fit the models without them. However, we will keep them in mind in subsequent models.

10. Fit a quasi-Poisson GLM to the Wong data, and look at the analysis of deviance table to examine the significance of the model terms. You need to use test = "F" with the anova command, because the corrected test statistics follow an F distribution rather than a χ^2 distribution.

```
> wong.glm.qpois = glm(Count ~ V * H * P, data = wong.df,
    family = quasipoisson)
> wong.glm.qpois1 = glm(Count ~ V + H + P, data = wong.df,
    family = quasipoisson)
> anova(wong.glm.qpois, test = "F")

Analysis of Deviance Table

Model: quasipoisson, link: log

Response: Count

Terms added sequentially (first to last)
```

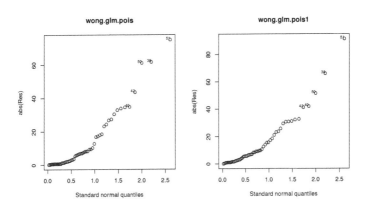

FIGURE 6.7: Half-normal plots for the residuals of the two Poisson GLM models for the Wong data

```
           Df Deviance Resid. Df Resid. Dev      F   Pr(>F)
NULL                         53     261782
V           1   205238       52      56543 320.47 < 2e-16
H           2    23377       50      33167  18.25 2.0e-06
P           1     3933       49      29234   6.14   0.017
V:H         2     3060       47      26174   2.39   0.104
V:P         1        6       46      26168   0.01   0.923
H:P         2      677       44      25491   0.53   0.593
V:H:P       2       14       42      25477   0.01   0.989

NULL
V      ***
H      ***
P      *
V:H
V:P
H:P
V:H:P
---
Signif. codes:
          0 '***' 0.001 '**' 0.01 '*' 0.05 '.' 0.1 ' ' 1
```

Note that the quasi-Poisson model only helps us in the tests of significance for the model terms. It makes no difference at all to the regression coefficients or the fitted values. It will, however, correct the standard errors in the regression summary table.

11. Fit the negative binomial model to the Wong data without the interaction terms. That is, fit a negative binomial GLM using an additive model.

    ```
    > library(MASS)
    > wong.glm.nb = glm.nb(Count ~ V + H + P, data = wong.df)
    ```

12. Create a half-normal plot of the residuals for the negative binomial model.

    ```
    > halfnorm(wong.glm.nb)
    ```

    This plot is shown in Figure 6.8. We can see from Figure 6.8 that points 38 and 51 stick out again. The half-Normal plot moves along in small steps and then jumps up at point 38 and 51. This is an indication that they might be outliers. Inspection of the data shows why point 51 is being flagged.

    ```
    > wong.df$Count
    ```

```
[1]    194    163    190    200    228    250    402    219
[9]    406    198    456    180    571    301    324    275
[17]   402    214   2062    615   2747    674    987    624
[25]  2361   1053   1692   1120   3712   1910   3940   1674
[33]  1989   1547   6002   1497   2552   7798   3013   2848
[41]  6014   4888   4793   3681   5127   8955   5027   5148
[49] 14496   5499  22035  13814  17659  14380
```

```
> wong.df$Count[51]
```

```
[1] 22035
```

With a value of 22,035 it is clearly the maximum of the data set and it has a count of about 4,500 more than the next closest observation. We need to do a little more digging (with the subset command) to find out what is interesting about point 38.

```
> wong.df[38, ]
```

```
   Count sqrt.count log.count    V     H  P
38  7798     88.306    8.9616 1147 HRH55 RN
```

```
> subset(wong.df, H == "HRH55" & P == "RN")
```

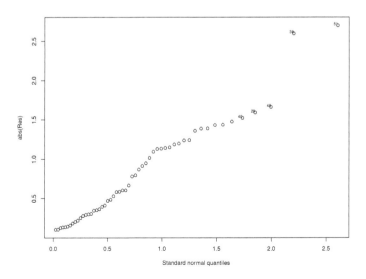

FIGURE 6.8: Half-normal plot for the residuals of the negative binomial GLM for the Wong data

```
    Count sqrt.count log.count    V    H   P
1     194     13.928   5.2679   574 HRH55 RN
3     190     13.784   5.2470   562 HRH55 RN
5     228     15.100   5.4293   585 HRH55 RN
19   2062     45.409   7.6314   919 HRH55 RN
21   2747     52.412   7.9183   987 HRH55 RN
23    987     31.417   6.8947   900 HRH55 RN
37   2552     50.517   7.8446  1087 HRH55 RN
38   7798     88.306   8.9616  1147 HRH55 RN
39   3013     54.891   8.0107  1107 HRH55 RN
46   8955     94.631   9.1000  1451 HRH55 RN
47   5027     70.901   8.5226  1303 HRH55 RN
48   5148     71.750   8.5464  1367 HRH55 RN

> subset(wong.df, H == "HRH55" & P == "RN" &
    V > 1050 & V < 1150)

    Count sqrt.count log.count    V    H   P
37   2552     50.517   7.8446  1087 HRH55 RN
38   7798     88.306   8.9616  1147 HRH55 RN
39   3013     54.891   8.0107  1107 HRH55 RN
```

If we restrict our examination to the three data points where the velocity is about 1100 fts^{-1}, then we can see that observation 38 has a Count value that is twice as high as the other two replicates in this set of experiments.

```
> wong.glm.nb1 = glm.nb(Count ~ V + H + P, subs = -c(38,
    51), data = wong.df)
```

We can compare the effect on the quality of the fit by using our goodness-of-fit measure for both models

```
> 1 - pchisq(deviance(wong.glm.nb), df.residual(wong.glm.nb))
```

[1] 0.26645

```
> 1 - pchisq(deviance(wong.glm.nb1), df.residual(wong.glm.nb1))
```

[1] 0.26843

Hardly an earth shattering difference!

13. Finally, we come to the question "Which model is the best?" There are a variety of different measures. We can compute another R^2 type statistic for each of the GLMs, which is

$$1 - \frac{D_{res}}{D_{null}}$$

where D_{res} is the residual deviance after all the terms have been fitted, and D_{null} is the null deviance. We can extract this information from the **summary** object that we get from applying the **summary** command to the fitted **glm** object. I have included code for a small function below that does just this. We could put the linear model on an equal footing with the other models by using fitting a Gaussian or normal GLM with the canonical identity link; however, the deviance for the normal GLM is exactly equal to the sum of squares—try it for yourself. The default value of family in the **glm** command is **gaussian**.

```
> wong.glm.norm = glm(log.count ~ V + P + H,
    data = wong.df)
> anova(wong.glm.norm)

Analysis of Deviance Table

Model: gaussian, link: identity

Response: log.count

Terms added sequentially (first to last)


     Df Deviance Resid. Df Resid. Dev
NULL                  53       109.9
V     1     96.3     52        13.6
P     1      2.0     51        11.6
H     2      6.0     49         5.6

> anova(wong.lm.log)

Analysis of Variance Table

Response: log.count
          Df Sum Sq Mean Sq F value   Pr(>F)
V          1   96.3    96.3   846.5 < 2e-16 ***
P          1    2.0     2.0    17.2 0.00013 ***
H          2    6.0     3.0    26.5 1.6e-08 ***
Residuals 49    5.6     0.1
---
Signif. codes:
         0 '***' 0.001 '**' 0.01 '*' 0.05 '.' 0.1 ' ' 1

> summary(wong.lm.log)$r.squared

[1] 0.94926
```

```
> dev.explained = function(fit) {
    round(1 - with(summary(fit), deviance/null.deviance),
        3)
}
> dev.explained(wong.glm.norm)
[1] 0.949

> dev.explained(wong.glm.pois1)
[1] 0.888

> dev.explained(wong.glm.qpois1)
[1] 0.888

> dev.explained(wong.glm.nb)
[1] 0.939
```

Using these measures we would probably decide that either the linear model with a log response or the negative binomial are the models that we should consider. The negative binomial model explicitly models count data, and for that reason I would argue that it is preferable.

6.6 Logistic regression or the binomial GLM

The binomial GLM is often referred to as the **logistic regression** model, and in fact many users are often unaware that they are using a GLM. Technically, the binomial GLM should be called the Bernoulli GLM, as it models the mean of a Bernoulli random variable. The Bernoulli distribution, named after Swiss scientist Jacob Bernoulli, is used to model data that only have two possible values: a success or a failure. A random experiment with only two possible outcomes is often called a **Bernoulli trial**. The mean of the Bernoulli distribution is equal to the probability of a success, p. A binomial random variable is constructed by adding up the results of n trials, where each success has value 1 and each failure value 0. The binomial GLM models a function of the probability of success. Specifically, if we have data Y_i, which we believe follow a binomial distribution, where the probability of success is affected by covariate(s) X_i, then we can model this situation using the binomial GLM. The model is

$$y_i \sim Bin(n_i, p_i)$$
$$\ln\left(\frac{p_i}{1-p_i}\right) = \beta_0 + \beta_1 x_{1i} + \beta_2 x_{2i} + \cdots + \beta_p x_{pi}$$

The function
$$\ln\left(\frac{p_i}{1-p_i}\right)$$
is called the **logit**. There are a number of justifications for its use. In the first instance you can see that it is simply the log odds that we used in Chapter 4. The second reason is that it is the inverse of the **logistic function**

$$f(z) = \frac{1}{1+\exp(-z)} = \frac{\exp(z)}{1+\exp(z)} \qquad (6.1)$$

The logistic function has a useful property that it takes any real value z, such that $-\infty < z < \infty$ and maps it to a value between 0 and 1.

This is useful, because it lets us translate probability, which is measured on the range $[0,1]$ onto the range $[-\infty, \infty]$ removes some constraints from the estimation process. Finally, we can also justify the logit function using statistical theory which shows that it is the canonical link for the binomial. Note that there are two alternative link functions that are used with the binomial GLM. These are the **probit** and the **complementary log-log** (usually abbreviated clog-log) link. Probit regression uses the probit link. However, the logit link is the most commonly used, and it is all we will cover in this book.

As noted earlier, the logit function can be thought of as the log odds. This relationship is useful because we can use logistic regression to analyze tabular data where one of the dimensions of the table is two. That is, if we have a

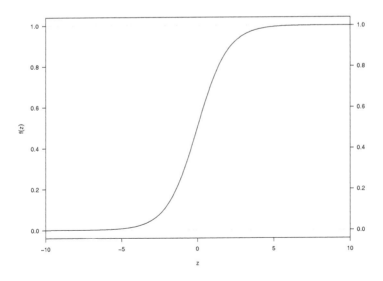

FIGURE 6.9: The logistic function maps $z \in [-\infty, \infty]$ to $[0,1]$

table with two rows or two columns, then we can use logistic regression to estimate the (log) odds.

6.6.1 Example 6.5—Logistic regression for SIDS risks

In Chapter 4 we analyzed the data of Törrő et al. [25]. The authors of this study were interested in the frequency of extramedullary haematopoiesis (EMH) in liver of sudden infant death (SIDS) cases. The data from this study are in the data frame liver.df in the dafs library. The data record information on the presence or absence of EMH symptoms in 51 SIDS and 102 non-SIDS cases. To use a binomial GLM, we let the response be the variable emh, which is a factor with two levels (no-emh and emh), and let the explanatory variable be sids, which has two levels (non-sids and sids). We fit the model using the glm function with the parameter family equal to binomial.

	Estimate	Std. Error	z value	$\Pr(>|Z|)$
(Intercept)	2.7726	0.4208	6.59	0.0000
sidssids	-1.0908	0.5704	-1.91	0.0558

TABLE 6.9: Summary table for the liver binomial GLM

Table 6.9 shows the regression summary table for the fitted model. There are a number of interesting features in this table. Firstly, the coefficient for sidssids gives the log odds ratio of EMH in the SIDS group. If we exponentiate it $\exp(-1.091) = 0.336$, then we can see that it is almost identical to the odds ratio estimated by the fisher.test (0.339).

Secondly, we can see that the regression coefficient is not significant at the 5% level. This means we believe that there is a chance that this result could have arisen from sampling variation alone. The $100(1-\alpha)$% confidence interval on the odds ratio can be easily calculated from

$$\exp(\widehat{\beta}_i \pm q(1-\alpha/2)se(\widehat{\beta}_i))$$

where $q(1-\alpha/2)$ is the $1-\alpha/2$ quantile from the standard normal distribution. It is important to note we are using quantiles from the normal distribution and not Student's t-distribution. Using our example data, we have $\widehat{\beta}_1 = -1.091$, $se(\widehat{\beta}_1) = 0.5704$ so a 95% confidence interval for the odds ratio is

$$\begin{aligned} CI &= \exp(-1.091 \pm 1.96 \times 0.5704) \\ &= \exp([-2.21, 0.03]) \\ &= [0.11, 1.03] \end{aligned}$$

> **Tip 19: Coefficients in logistic regression**
>
> The coefficient on an explanatory variable, x, in a logistic regression model gives the logarithm of the **odds ratio**, not the odds, for a 1-unit change in the x variable. This is because the difference to two logits is equal to the odds ratio. If we have two sets of odds O_1 and O_2, then
>
> $$O_1 = \frac{p_1}{1-p_1} \text{ and } O_2 = \frac{p_2}{1-p_2}$$
>
> for some probabilities p_1 and p_2. The ratio of these two sets of odds is then
>
> $$\frac{O_1}{O_2} = \frac{p_1}{1-p_1} \div \frac{p_2}{1-p_2}$$
>
> If we take the logarithm of both sides of the last equation, we get
>
> $$\log\left(\frac{O_1}{O_2}\right) = \log\left(\frac{p_1}{1-p_1} \div \frac{p_2}{1-p_2}\right)$$
> $$= \text{logit}(p_1) - \text{logit}(p_2)$$

We can see that this is a slightly tighter interval than the interval from `fisher.test`. However, the conclusion remains the same. The confidence interval includes one, and therefore we think there no or weak evidence of elevated probability (occurrence) of EMH in the SIDS compared to the non-SIDS groups.

6.6.2 Logistic regression with quantitative explanatory variables

There is a long history of the application of logistic regression to **contingency tables**. That is, tables that are constructed by the cross-classification of a sample or samples on one or more qualitative (factor) variables. However, there is no requirement that the explanatory variables in logistic regression be factors. There is no extra effort required in fitting a model with one or more quantitative explanatory variables. Similarly, we can have a mix of quantitative and qualitative (factor) variables.

6.6.3 Example 6.6—Carbohydrate deficient transferrin as a predictor of alcohol abuse

Berkowicz et al. [51] were interested in the stability of carbohydrate-deficient transferrin concentration in vitreous humour (VH-CDT). VH-CDT has been shown to be useful for a diagnosis of alcohol misuse. Berkowicz et al. measured the VH-CDT concentration in 21 alcoholics and 7 non-alcoholics.

The data for each group are shown in Figure 6.10. We can see from Figure 6.10 that on average the VH-CDT concentration is higher in the alcoholic group than the non-alcoholic group, but there is some overlap. We might use logistic regression to model the probability of a diagnosis of alcoholism on the basis of the VH-CDT concentration. In this example, the probability of a "success"—i.e., a diagnosis of alcoholism, changes continuously with respect to the value of VH-CDT. We fit the model using the `glm` command with `family` equal to `binomial`. We can see from the summary table of the logistic

	Estimate	Std. Error	z value	$\Pr(>|Z|)$
(Intercept)	-3.4091	1.5943	-2.14	0.0325
vhcdt1	0.8272	0.2964	2.79	0.0053

TABLE 6.10: Summary table for logistic regression of alcoholism status on VH-CDT concentration

regression analysis (Table 6.10) that VH-CDT does indeed play an important role in predicting alcohol abuse status. We interpret the regression as a 1 unit change in VH-CDT concentration leads to the odds on a diagnosis of alcohol abuse almost doubling (2.3 times). Logistic regression analysis is often carried out with one of two purposes. The first, and possibly the most common, is the association of explanatory variables with changes in the response. That is, the significance of the explanatory variables is of importance. The second purpose

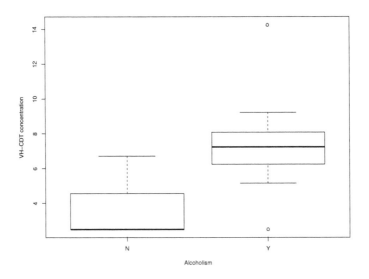

FIGURE 6.10: Mean VH-CDT concentration by alcoholism status

is the construction of a model for classification. A logistic regression model, given certain values of the explanatory variables, will return the (transformed) probability that the observation is a "success." If we use this probability with a **classification rule**, then we can classify observations as being either a "success" or a "failure." That is, we can use a logistic regression model to assign observations to one of two different classes. This process is called **classification**.

The fitted probabilities for the VH-CDT model are shown in Figure 6.11. If we impose an arbitrary classification rule where we return a diagnosis of alcoholism if the probability given the VH-CDT concentration is greater than 0.5, then we can see that we would misclassify several people. It is a little hard to see from the plot. A **confusion matrix** is a visualization tool from the fields of artificial intelligence and data mining, which can be used to assess the performance of a classifier. The confusion matrix, as the name suggests, is a matrix with two rows and two columns. The columns of the matrix represent true class memberships the data. The rows of the matrix represent the predicted class membership. If we let $Positive = Success$ and $Negative = Failure$, then the layout of a confusion matrix is shown in Table 6.11. If a model or technique classifies the data perfectly, then the number of false positives and false negatives will be zero.

The confusion matrix for our data is shown in Table 6.12. We can see that we have two false positives and one false negative. That is, we have two individuals who are non-alcoholic classified as alcoholic and one alcoholic

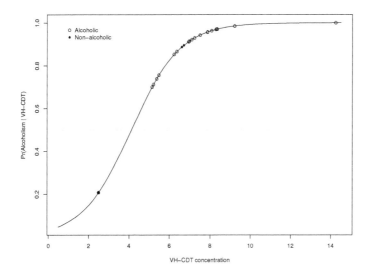

FIGURE 6.11: The fitted probability curve for the VH-CDT data

		Actual	
Predicted		Positive	Negative
	Positive	True Positive	False Positive
	Negative	False Negative	True Negative

TABLE 6.11: A hypothetical confusion matrix

		Actual	
Predicted		Y	N
	Y	20	2
	N	1	5

TABLE 6.12: Confusion matrix for the logistic model

classified as non-alcoholic. Using the data you *trained* the classifier/model with to test the prediction performance is not strictly kosher. However, this is not a book on machine learning or data mining. The false positive rate for our example is $2/27 \approx 7.4\%$. The false negative rate is $1/27 \approx 3.7\%$, and the overall misclassification rate or error rate is $3/27 \approx 11.1\%$.

6.6.4 Example 6.7—Morphine concentration ratios as a predictor of acute morphine deaths

Levine et al. [52] were interested in whether comparison of morphine concentrations from a central and peripheral site could be used to determine whether a morphine death was acute or delayed. They classified 126 cases where morphine was involved as either acute or random. The acute cases were those where the urine-free morphine concentration was less than 25 ng/mL. In each case the morphine concentration was determined in from two samples: a heart blood sample and a peripheral blood sample. The theory was that the ratio of these two samples would help predict *acute* opiate intoxication deaths versus *delayed* deaths when the only information available is heart and peripheral blood-free morphine concentrations. In this example we have two possible outcomes *acute* and *delayed* (or *random*). We have two continuous explanatory variables which will be correlated with each other because they are measured on the same individual, but this is overcome by the fact that it is the ratio of these two variables, not the variables themselves, is of interest. Ratio variables tend to not be very well behaved, so for that reason, we will use the log-ratio rather than the ratio itself. As we have only one explanatory variable, a box plot of the these values by outcome is useful. We can see from Figure 6.12 that there is considerable overlap between the two groups. We would predict therefore, that the log-ratio is probably not going to be very useful in predicting acute cases versus delayed cases. We can test this idea explicitly using logistic regression. Our model is

$$Y_i|X_i = \log\left(\frac{H_i}{P_i}\right) \sim Bernoulli(\pi_i)$$

$$E\left[Y_i|X_i = \log\left(\frac{H_i}{P_i}\right)\right] = \pi_i$$

$$\text{logit}(\pi_i) = \beta_0 + \beta_1 \log\left(\frac{H_i}{P_i}\right)$$

where H_i and P_i are the concentration of morphine in the heart and peripheral blood samples of the ith case. This is another way of writing the model for a binomial GLM. We can see from Table 6.13 that the log-ratio of morphine

	Estimate	Std. Error	z value	Pr(> |Z|)
(Intercept)	0.5127	0.1949	2.63	0.0085
log.ratio	-0.6814	0.4313	-1.58	0.1142

TABLE 6.13: Regression summary for the morphine concentration data

concentration is not significant at the 0.05 level or even at the 0.1 level. The

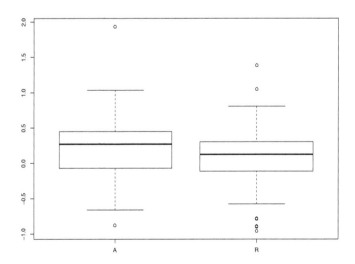

FIGURE 6.12: Plot of the log-ratio of morphine concentration for acute (A) and random (R) cases

	Female				Male			
	blunt	comb	dist	seatbelt	blunt	comb	dist	seatbelt
non	2	3	2	5	16	8	4	14
severe	4	3	6	2	19	0	0	5

TABLE 6.14: Cross classification of the experiments by `gender`, `injury`, and `load.cond`

coefficient on the log-ratio is −0.68, so the odds ratio is 0.51. The odds ratio is for the random group, not the acute group because R recodes the base level of the factor as 0 and the next level as 1. Therefore, this coefficient is a reduction on the odds of being a random case. That is, as the log ratio goes up the odds ratio that the case is random goes down. A 95% confidence interval is given by

$$C.I. = \exp\left(\widehat{\beta}_1 \pm 1.96 \times se(\widehat{\beta}_1)\right)$$
$$= \exp\left(-0.6814 \pm 1.96 \times 0.4313\right)$$
$$= (0.22, 1.18)$$

6.6.5 Example 6.8—Risk factors for thoracic injuries

Kent and Patrie [49] analyzed their crash test data using logistic regression. To do this they recoded the response variable which recorded the number of rib fractures in a variety of different ways. In this example we will use Kent and Patrie's classification of greater than six rib fractures as *severe injury* and six or fewer fractures as *no severe injury*. We will code these two outcomes in R as a new variable `injury` with possible values `severe` and `non`. The data set contains four possible explanatory variables, which are described in Section 6.5.2 (p. 223). We start by fitting the saturated model. When we do this R warns us that some of the fitted probabilities are zero or one. We discuss this problem in the next section. One potential pitfall is that there are some "cells" with zero counts. We can think of the categorical explanatory variables as cross-classifying the response. If the categorical variables have too many categories, then some of the combinations of categories will have no cases. This is part of the problem in this data set. If we cross-classify by `injury`, `gender` and `load.cond`, we find that there are no males with severe injuries in the combined and distributed categories for `load.cond`. This cross-tabulation is shown in Table 6.14. We can overcome this by not fitting interaction terms involving both `gender` and `load.cond`. Therefore, we will fit the model that Kent and Patrie claim has a biomechanical justification, which is

$$\text{logit}(\Pr(injury = severe)) \sim cmax + load.cond + gender * age$$

The analysis of deviance table for this model is shown in Table 6.15. We can

	Df	Deviance	Resid. Df	Resid. Dev.	Pr($> \chi^2$)
NULL			92	126.50	
cmax	1	38.94	91	87.55	0.0000
age	1	5.22	90	82.34	0.0223
gender	1	1.29	89	81.04	0.2553
load.cond	3	0.77	86	80.28	0.8576
age:gender	1	2.82	85	77.46	0.0932

TABLE 6.15: A logistic regression model for injury level in the crash data

see that cmax and age are significant in this model. This reduced model has reasonable fit by our χ^2 criterion, as $1 - Pr(\chi^2_{90} > 82.34) = 0.7$.

6.6.6 Pitfalls for the unwary

One of the strange limitations of logistic regression is that it does not work when the explanatory variables **completely separates** or perfectly predicts the response. That is, if our explanatory variable measures dose for example, and all of the successes ($y = 1$) have dose with values greater than 10, and all of the failures ($y = 0$) have dose with values less than 10, then the variable dose completely separates the response. This is problematic because of the way the parameters in logistic regression are estimated. The problem is best explained graphically. Figure 6.13 shows the behavior of the logistic function as the overlap (in x-values) between the two groups decreases. The figure in the left-hand corner of each plot is the percentage of (theoretical) overlap between the ranges of the two groups. We can see that the logistic curve gets *steeper* as the overlap between the two groups decreases. This steepness is controlled by the coefficient on x, β_1. When there is no overlap between the two groups, the value of β_1 is infinite (in magnitude). This causes the maximum likelihood procedure, which is used to estimate the coefficients, to break down.

Complete separation is usually relatively easy to detect when we only have one or two explanatory variables, or when one explanatory variable alone predicts the response perfectly. However, this problem also can occur when we get close to *near separation*. Near separation occurs when the data are almost perfectly separated except for one or two cases. R will warn you that this may be happening by telling you that some of the probabilities are estimated to be near 0 or 1. This happens when one or more of the regression coefficients become moderately large in magnitude.

244   *Introduction to data analysis with R for forensic scientists*

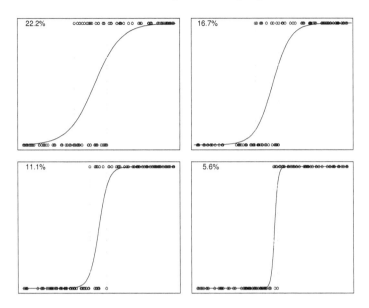

FIGURE 6.13: The behavior of the logistic function as group overlap decreases

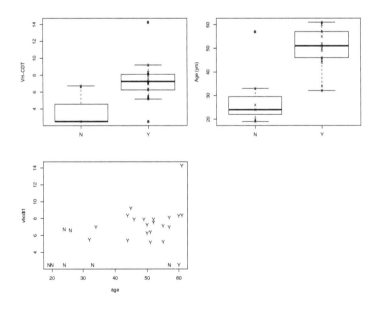

FIGURE 6.14: Response (alc) with respect to Age and VH-CDT concentration

6.6.7 Example 6.9—Complete separation of the response in logistic regression

Figure 6.14 shows plots of the response alc in the VH-CDT data set with respect to age and vhcdt1. You can see that either variable provides quite good separation, and when the two are combined, only two cases are misclassified. Those of you who are you trying to replicate the line in Figure 6.14 should know that it was fitted very unscientifically "by eye." No doubt someone could (and will) find a better line with a support vector machine (SVM) classifier. When we try and fit a logistic regression model with both of these variables we get a warning message. This is because the linear combination of vhcdt1 and age almost perfectly separates the response.

6.6.8 Tutorial

1. Load the liver data.

    ```
    > data(liver.df)
    > names(liver.df)

    [1] "sids" "emh"
    ```

2. Fit a logistic regression (binomial GLM) model with emh as the response and sids as the explanatory variable.

    ```
    > liver.fit = glm(emh ~ sids, data = liver.df,
        family = binomial)
    ```

3. Test the significance of the variable sids by looking at the χ^2-test from the analysis of deviance table.

    ```
    > anova(liver.fit, test = "Chisq")

    Analysis of Deviance Table

    Model: binomial, link: logit

    Response: emh

    Terms added sequentially (first to last)


          Df Deviance Resid. Df Resid. Dev P(>|Chi|)
    NULL                   152       93.6
    sids   1     3.69      151       90.0     0.055 .
    ---
    Signif. codes:  0 '***' 0.001 '**' 0.01 '*' 0.05 '.' 0.1 ' ' 1
    ```

4. Examine the model summary table.

```
> summary(liver.fit)

Call:
glm(formula = emh ~ sids, family = binomial, data = liver.df)

Deviance Residuals:
    Min      1Q  Median      3Q     Max
 -2.380   0.348   0.348   0.584   0.584

Coefficients:
             Estimate Std. Error z value Pr(>|z|)
(Intercept)    2.773      0.421    6.59  4.4e-11 ***
sidssids      -1.091      0.570   -1.91    0.056 .
---
Signif. codes:  0 '***' 0.001 '**' 0.01 '*' 0.05 '.' 0.1 ' ' 1

(Dispersion parameter for binomial family taken to be 1)

    Null deviance: 93.637  on 152  degrees of freedom
Residual deviance: 89.950  on 151  degrees of freedom
AIC: 93.95

Number of Fisher Scoring iterations: 5
```

Note that the P-value for sids is slightly different than the P-value from the analysis of deviance table. In general, although there is an order dependence, the test from the analysis of deviance table will be more accurate.

5. Get an estimate for the odds ratio for SIDS and a confidence interval for the odds ratio. To do this we need to extract the regression table from the summary output. The steps to do this are listed below. Basically we

   (i) Store the information returned from summary
   (ii) Extract the coefficient table using $coefficients
   (iii) Take the second row of the coefficient table which is the row for the sids variable

```
> summary.table = summary(liver.fit)
> coef.table = summary.table$coefficients
> b1 = coef.table[2, ]
> b1
```

Modeling count and proportion data

```
Estimate Std. Error    z value  Pr(>|z|)
-1.090830   0.570381  -1.912458  0.055818
```

The estimate can be obtained from the 1st element of b1, and the standard error from the 2nd element of b1. We construct a confidence interval using appropriate percentage point from the normal distribution. That is,

```
> est = b1[1]
> se = b1[2]
> ci = est + c(-1, 1) * qnorm(0.975) * se
> ci = exp(ci)
> ci

[1] 0.10984 1.02747
```

6. Load the VH-CDT data.

```
> data(vhcdt.df)
> names(vhcdt.df)

[1] "age"    "alc"    "vhcdt1" "vhcdt2" "vhtf"
[6] "td"     "th"
```

7. Produce a box plot of VH-CDT concentration by alcohol abuse status, and overlay the box plot with the data points.

```
> boxplot(vhcdt1 ~ alc, data = vhcdt.df)
> points(rep(2:1, c(21, 7)), vhcdt.df$vhcdt1,
     pch = "x")
```

8. Fit a logistic regression model with alc as the response and vhcdt1 as the explanatory variable.

```
> vhcdt.fit = glm(alc ~ vhcdt1, data = vhcdt.df,
     family = binomial)
> anova(vhcdt.fit, test = "Chisq")

Analysis of Deviance Table

Model: binomial, link: logit

Response: alc

Terms added sequentially (first to last)


     Df Deviance Resid. Df Resid. Dev P(>|Chi|)
```

248   *Introduction to data analysis with R for forensic scientists*

```
NULL                       27      31.5
vhcdt1     1     12.6      26      18.8     0.00038 ***
---
Signif. codes:
               0 '***' 0.001 '**' 0.01 '*' 0.05 '.' 0.1 ' ' 1
```

9. We can calculate some measures of model fit in the same way we did for the Poisson model.

```
> 1 - pchisq(deviance(vhcdt.fit), df.residual(vhcdt.fit))

[1] 0.84314

> null.dev = anova(vhcdt.fit)$"Resid. Dev"[1]
> 100 * (1 - deviance(vhcdt.fit)/null.dev)

[1] 40.175
```

10. The fitted function returns the probability of alcoholism conditional on the VH-CDT concentration. We can use the ifelse function to recode the probabilities into predicted diagnoses. To do this we need a classification rule. We will choose a simple "probability > 0.5" (sometimes called *preponderance of the evidence* or *balance of probabilities* in legal circles). That is, if the probability is greater than 0.5, then we will code the diagnosis as Y for yes. If the probability is less than (or equal to) 0.5, then we will code it N for no.

```
> vhcdt.pred = ifelse(fitted(vhcdt.fit) > 0.5,
      "Y", "N")
```

11. Use the predicted values to create a confusion matrix.

```
> vhcdt.df$alc = factor(vhcdt.df$alc,
                    levels = c("Y", "N"))
> vhcdt.pred =  factor(vhcdt.pred,
                    levels = c("Y", "N"))
> conf.matrix = xtabs(~vhcdt.pred+vhcdt.df$alc)
> conf.matrix

              vhcdt.df$alc
vhcdt.pred  Y   N
         Y 20   2
         N  1   5
```

The first two lines of the code above are not absolutely necessary. All they do is make sure that R orders the levels of the factor so that Y is followed by N rather than the default alphabetical order.

12. Fit the logistic regression model using both vhcdt1 and age as predictors. We know that this will give us problems, but it is a useful exercise.

```
> vhcdt.fit1 = glm(alc ~ vhcdt1 + age, data = vhcdt.df,
    family = binomial)
```

Note the warnings that occurred. If we ignore these and go ahead and look at some of the output, we get confusing information.

```
> anova(vhcdt.fit1, test = "Chisq")

Analysis of Deviance Table

Model: binomial, link: logit

Response: alc

Terms added sequentially (first to last)


        Df Deviance Resid. Df Resid. Dev P(>|Chi|)
NULL                    27       31.49
vhcdt1   1    12.6     26       18.84    0.00038 ***
age      1    12.1     25        6.77    0.00051 ***
---
Signif. codes:
       0 '***' 0.001 '**' 0.01 '*' 0.05 '.' 0.1 ' ' 1

> summary(vhcdt.fit1)

Call:
glm(formula = alc ~ vhcdt1 + age, family = binomial,
       data = vhcdt.df)

Deviance Residuals:
      Min        1Q    Median         3Q        Max
-1.770842 -0.023380 -0.003271   0.000232   1.154969

Coefficients:
             Estimate Std. Error z value Pr(>|z|)
(Intercept)   28.923     23.643    1.22    0.22
vhcdt1        -2.785      2.412   -1.15    0.25
age           -0.384      0.302   -1.27    0.20

(Dispersion parameter for binomial family taken to be 1)

    Null deviance: 31.491  on 27  degrees of freedom
```

```
Residual deviance:  6.771  on 25  degrees of freedom
AIC: 12.77

Number of Fisher Scoring iterations: 10
```

The analysis of deviance table tells us that both variables are important in predicting the response, which they are, but the summary table tells us that neither variable is significant. We know that in this situation the computer has difficulty estimating the parameters. This is reflected somewhat in the estimates of the standard errors of the regression coefficients. At any rate, the warning message should be taken as a signal that perhaps this is not the best model. These separation problems can sometimes be overcome by using the bias-reduced estimates proposed by Firth [53], which are implemented in the R package brglm [54].

13. **Advanced** You may recall that measurements of VH-CDT, which fell below the limit of detection (LOD), were arbitrarily recoded to have a value of 2.5 µg/L. We can examine the effect of this recoding by changing these values to uniform random numbers between 0 and 5, and re-fitting the model. Of course, any set of numbers we choose would be just as arbitrary as setting them all to 2.5. Therefore, we avoid settling for just one set of numbers and instead randomly select N sets. Each set will give a slightly different fitted model, and hence slightly different fitted values. If N is sufficiently large, then we can get some idea of the variability in the fitted values for these missing observations.

   (i) We will use $N = 1,000$. It is a good compromise between accuracy and computing time. We will set up space to store $N = 1,000$ sets of $n = 28$ fitted values.
   ```
   > N = 1000
   > fitted.values = matrix(0, nc = 28, nrow = N)
   ```

   (ii) Next, we will copy the variables alc and vhcdt1 to new variable y and x. We will also determine which values are missing (those that have x==2.5) using the which function and how many values are missing using the length function.
   ```
   > data(vhcdt.df)
   > x = vhcdt.df$vhcdt1
   > y = vhcdt.df$alc
   > miss = which(x == 2.5)
   > nmiss = length(miss)
   ```

   (iii) The remainder of the process is iterative. On each step we select a new set of values to replace the missing values from a $Uniform(0,5)$ distribution, re-fit the logistic regression model, and store the fitted values in a row of the fitted.values matrix. A $Uniform(0,5)$

distribution is one where every real value between 0 and 5 is equally likely to occur.

```
> for (i in 1:N) {
    x[miss] = runif(nmiss, 0, 5)
    fit = glm(y ~ x, family = binomial)
    fitted.values[i, ] = fitted(fit)
}
```

(iv) Finally, we can plot the results. I will use the errbar function from the Hmisc package to plot the median probability, and 2.5% and 97.5% percentiles.

```
> library(Hmisc)
> meds = apply(fitted.values, 2, median)
> uq = apply(fitted.values, 2, quantile, probs = 0.975)
> lq = apply(fitted.values, 2, quantile, probs = 0.025)
> errbar(1:28, meds, uq, lq, col = ifelse(y ==
    "Y", "red", "blue"), xlab = "VH-CDT",
    ylab = "Probability")
```

Figure 6.15 shows the resulting graph. We can see that there is quite a bit of uncertainty for the missing values, as there should be, and that a classification rule of $P > 0.5$ would not give us such good classification.

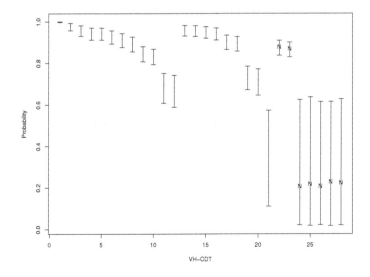

FIGURE 6.15: Probability of alcohol abuse with sensitivity to LOD analysis limits

14. Load the morphine concentration data

    ```
    > data(morphine.df)
    > names(morphine.df)

    [1] "subject" "loc"     "type"    "conc"

    > levels(morphine.df$type)

    [1] "A" "R"

    > levels(morphine.df$loc)

    [1] "H" "P"
    ```

15. To perform the analysis, we need to reorganize the data so that we can create a log ratio variable. To do this we need to separate out the peripheral blood concentration measurements and append ratio as a separate column in a new data frame.

    ```
    > idx = which(morphine.df$loc == "H")
    > concP = morphine.df$conc[-idx]
    > morphine.df = morphine.df[idx, ]
    > morphine.df = within(morphine.df, {
         log.ratio = log(conc/concP)
      })
    ```

16. Produce a box plot of the log ratio concentration by case type.

    ```
    > boxplot(log.ratio ~ type, data = morphine.df)
    ```

17. Now we can fit the logistic regression model.

    ```
    > morphine.fit = glm(type ~ log.ratio, data = morphine.df,
         family = binomial)
    > anova(morphine.fit, test = "Chisq")

    Analysis of Deviance Table

    Model: binomial, link: logit

    Response: type

    Terms added sequentially (first to last)


              Df Deviance Resid. Df Resid. Dev P(>|Chi|)
    NULL                       125        169
    log.ratio  1     2.62      124        167      0.11
    ```

We can see from the analysis of deviance table that the log ratio is not really doing much at all in terms of aiding the prediction of acute cases.

6.7 Deviance

I will define deviance without trying to explain it too much. We need some terminology and definitions. The first of these is **likelihood**.

If $\vec{x} = \{x_1, x_2, \ldots, x_n\}$ is a sample of size n from a distribution with probability function density $f(.)$, which depends on parameters $\vec{\theta} = (\theta_1, \theta_2, \ldots, \theta_k)$, then the **likelihood** of $\vec{\theta}$ given $\vec{x}$ is

$$L\left(\vec{\theta}|\vec{x}\right) = \prod_{i=1}^{n} f(x_i; \vec{\theta})$$

Implicit in this definition is that $x_1, x_2, \ldots, x_n$ are independent. It is very common in statistics to use the **log-likelihood**. This is often denoted $l(\vec{\theta}|\vec{x})$ or sometimes just $l(\vec{\theta})$ and is defined as

$$l(\vec{\theta}; \vec{x}) = \log L\left(\vec{\theta}|\vec{x}\right) = \sum_{i=1}^{n} \log\left(f(x_i; \vec{\theta})\right)$$

For example, $\vec{x} = (1, 1, 1, 1, 1, 1, 0, 0, 0, 1)$ is a sample of size $n = 10$ from a Bernoulli distribution. The pdf for a Bernoulli random variable is

$$f(x; \pi) = \begin{cases} \pi & , x = 1 \\ (1 - \pi) & , x = 0 \end{cases}$$

Therefore, the likelihood of π given the sample is

$$L(\pi|\vec{x}) = \pi \times \pi \times \pi \times \pi \times \pi \times \pi \times \pi \times (1 - \pi) \times (1 - \pi) \times (1 - \pi) \times \pi$$
$$= \pi^7 (1 - \pi)^3$$

and the log-likelihood is

$$l(\pi|\vec{x}) = 7 \log(\pi) + 3 \log(1 - \pi)$$

The likelihood function is useful because it provides us with a mechanism for talking about the parameters that best describe the data. The parameter values that maximize the likelihood are called the **maximum likelihood estimates**, or **MLEs**, for short. The process of choosing these values is called **maximum likelihood estimation**. In simple cases most MLEs correspond to the standard sample statistics we have already seen. For example, the sample mean $\bar{x}$ is the MLE for the population mean μ if the data are normally distributed, the sample proportion $\hat{\pi} = x/n$ is the MLE for the population proportion if the data are binomially (or Bernoulli) distributed and so on. In our example we can show graphically (Figure 6.16) that the likelihood is maximized when $\hat{\pi} = 0.7$, which we can easily see is the sample proportion of 7 successes in 10 trials. We can obtain the same result using calculus.

Generalized linear models are fitted using maximum likelihood estimation usually in conjunction with another process known as **iteratively reweighted least squares**, or **IRLS** for short. In this procedure, the estimates of the model parameters (the regression coefficients and variance) are chosen so that the likelihood is maximized. The difference between what we have seen so far and this situation is that the mean of the pdf changes with respect to the values of the explanatory variables. For example, if we have a simple linear regression model where

$$y_i \sim N(\beta_0 + \beta_1 x_i, \sigma^2), \quad i = 1, \ldots, n$$

then the likelihood for β_0, β_1, and σ given (y_i, x_i) is

$$L(\beta_0, \beta_1, \sigma) = \prod_i \varphi\left(\frac{y_i - (\beta_0 + \beta_1 x_i)}{\sigma}\right)$$

where $\varphi(.)$ is the pdf for the standard normal distribution.

If we have two models, m_1 and m_2, then we can calculate the deviance

$$D(y; m_1, m_2) = 2[l(y; m_1) - l(y; m_2)] \tag{6.2}$$

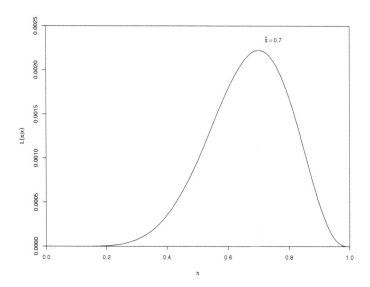

FIGURE 6.16: The likelihood for π given our Bernoulli data

Modeling count and proportion data 255

where $l(y|m_i)$ is the log-likelihood of the data y given model m_i. Usually m_1 is the **maximal model**. The definition of this term changes for each GLM. The maximal model for the Poisson GLM is found by setting $\lambda_i = y_i$, that is, the estimated means are the observed counts. The maximal model for the Binomial GLM is found by setting

$$\pi_i = \hat{\pi} = \frac{\sum_{i=1}^{n} I(y_i = 1)}{n}$$

where $I(y_i = 1)$ is an **indicator function** such that

$$I(y_i = 1) = \begin{cases} 1 & , y_i = 1 \\ 0 & , \text{otherwise} \end{cases}$$

That is, we set π equal to the proportion of successes in the response. This sounds terribly complicated; however, in practice it is not. We can treat the deviance in the same way as we treated the sums of squares at the end of the last chapter. That is, as terms are sequentially added they decrease the deviance. We start by adding a constant and then the model variables one by one. If the change in deviance is larger than what we would expect by random chance alone (according to a χ^2-test), then we take that as evidence that the term is useful in explaining the response.

The deviance information is given to us in the **Analysis of Deviance** table. This is sometimes called the ANODEV table. We get the analysis of deviance table by using the anova command on the fitted glm object. For our glass-breaking experiment example, we type

```
> anova(wong.fit, test = "Chisq")

Analysis of Deviance Table

Model: poisson, link: log

Response: Count

Terms added sequentially (first to last)


      Df Deviance Resid. Df Resid. Dev P(>|Chi|)
NULL                    53      261782
V      1   205238       52       56543    <2e-16 ***
H      2    23377       50       33167    <2e-16 ***
P      1     3933       49       29234    <2e-16 ***
---
Signif. codes:  0 '***' 0.001 '**' 0.01 '*' 0.05 '.' 0.1 ' ' 1
```

The first line of the table gives the deviance for what is termed the **null** model. The null model is the "constant only" model. That is, the null model is a model where the response is not affected by the explanatory variables at all. We have seen this before in simple regression. The null model is the intercept only model—the line is horizontal to the x-axis regardless of the value of x. The null model for the Poisson is $Y_i \sim Poisson(\lambda)$ where λ is unaffected by the explanatory variables. The null model for the binomial is $Y\ Binomial(n, \pi)$ where π is unaffected by the explanatory variables. You can think of the deviance for the null model as being analogous to the total sum of squares. The next line in the table describes the addition of the continuous variable V. The deviance decreases by approximately $205,000$. This is an absolutely massive change relative to the null deviance. The change in deviance is compared to the critical value for a χ^2 distribution with one degree of freedom. We take this degrees of freedom value from the Df column. The 95% critical value for a χ_1^2 distribution is 3.84. We continue in similar fashion. The addition of the two parameters ($df = 2$) needed to describe the three level factor H result in a change of deviance of approximately 23,000, which is very large compared to $\chi_2^2(0.95) = 5.99$. The addition of the single parameter needed to describe the two level factor P results in a change of deviance of approximately 4,000, which once again is very large compared to $\chi_1^2(0.95) = 3.84$. Therefore, we deem all three variables to be important in predicting the number of fragments of glass found on the ground.

Chapter 7

The design of experiments

No aphorism is more frequently repeated in connection with field trials, than that we must ask Nature few questions, or, ideally, one question, at a time. The writer is convinced that this view is wholly mistaken. Nature, he suggests, will best respond to a logical and carefully thought out questionnaire; indeed, if we ask her a single question, she will often refuse to answer until some other topic has been discussed – R. A. Fisher [55].

7.1 Introduction

This chapter provides an introduction to the statistical design and analysis of experiments. Initially, it might seem like an odd fit with the rest of the book. However, my motivation for writing this chapter came from my (anecdotal) observation that many of the experiments I have seen in forensic journals, books, and internal reports seemed to be curtailed in their scope. That is, the scientists and researchers involved limited the extent of their experiments because of physical constraints on resources. I felt that if those researchers knew about the field of experimental design then they could have done so much more with the resources they had to hand. I am also guilty of the same thing myself. Too often I have designed experiments with a resource hungry simple factorial structure when I could have achieved much more by designing a better, more efficient experiment. In my defence I will plead ignorance. My eyes were opened to the possible gains of good experimental design by teaching a third-year university course to statistics students. This forced me not only to revise what I had learned myself many years ago but to learn it to a much better level than I had as a student. As a consequence I now look at every experiment in terms of its design and whether that design took advantage of the features of the data. I hope I can convey some of that critical thinking in this chapter. Most books on experimental design are written for statisticians, and moreover statisticians without computers. There is an excessive body of literature on efficient ways to perform hand computation without much insight into analysis. Furthermore, this literature tends to focus on mathematically tractable problems. There is often a large gap between the theory of linear algebra (on which much of statistics depends) and the computational linear algebra used to compute estimates of the parameters and other quantities. I want to give practical advice in this chapter without spending very much time either on the mathematics or the computing complexities. Unfortunately, unlike previous chapters, it is not possible to offer example analyses for every situation. This is primarily because the data either do not exist or is nearly impossible to obtain. I will provide examples where I can, but I hope you the reader will forgive me if I cannot do so in every case. I also hope you can see some of the advantages in using the designs I describe. This will mean that perhaps in a future edition the data will exist.

I need to acknowledge my colleagues Chris Triggs and Arden Miller at the University of Auckland in this section, from whose excellent notes I have borrowed material.

7.2 Who should read this chapter?

This chapter is aimed at anyone planning to carry out an experiment. Even if you end up using the simplest design, I firmly believe that you will benefit from reading about the principles of experimental design and will gain valuable insight into the analysis of the resulting data.

7.3 What is an experiment?

In general, experiments are used to explore the behavior of systems or processes. As researchers we may have a belief that one or more factors will affect some quantity or feature that we can measure. An experiment involves varying those factors and assessing the effects.

What makes an experiment different from other situations is that the measurements are made under conditions that are directly controlled by the experimenter. It is this direct control that allows us to make statements about causation. For example, if we are in control of the application of water (or not) to a set of seedlings, then we can state that water makes the seedlings grow (or alternatively lack of water causes the plants to die.) This is different from an **observational study** like a survey. This distinction sometimes seems a little arbitrary to newcomers to statistics. The confusion arises because many of the same statistical techniques and models are used to analyze data from both designed experiments and observational studies. However, we should try to adhere to the rule that we can talk about *causation* in well-designed controlled experiments and *association* in observational studies.

7.4 The components of an experiment

Every experiment, regardless of the situation, has the same components.

7.4.1 Questions of interest?

Initially, we start with one or more objectives or questions of interest. These questions are our motivation to perform the experiment. Some examples from the data we have seen in previous chapters are

- Is there systematic spatial variation in the refractive index across a pane of glass?

- Is the number of fragments of glass found on the ground affected by the velocity, shape, and hardness of the bullet fired to break the window?

- Does the shotgun brand affect the pellet scatter pattern?

- Does the drinking container/beverage combination have an effect on the amount of DNA left?

7.4.2 Response variables

The process of defining the objectives of the experiment or the questions of interest leads to the question of response variables. One of the harder tasks in an experiment is deciding what an appropriate response variable or measurement is. An ideal response variable is one that

1. is strongly related to the question of interest

2. has low natural variability

3. has some sensitivity to changes in the treatment

4. is not expensive

Finding a response variable that is strongly related to the question of interest can be quite a contentious process in science. Care should be taken to show that the measurements relate to the question of interest in a sensible way. Just what "sensible" means of course is open for debate.

In our examples the response variable is defined directly by the question of interest:

- spatial variability experiment: the refractive indices of the glass fragments
- shotgun experiment: the area of the shot pattern
- drinking experiment: the amount of DNA left on the drinking vessel
- glass-breaking experiment: the number of glass fragments recovered from the ground

7.4.3 Treatment factors

The treatment factors, or sometimes just the treatments or the factors, are the inputs or variables that can be directly varied by the experimenter. We use the word **treatment** to indicate that these factors are under the control of the experimenter.

The treatment factors are things that the experimenter believes could affect the response. In our examples the treatment factors are

- Shotgun experiment:
  - distance from the target
  - shotgun brand
- Drinking container experiment:
  - drinking container/beverage combination
  - elapsed time between drinking and sample collection
- Glass-breaking experiment (Wong):
  - load/velocity of the projectile
  - shape of the projectile
  - hardness of the projectile

Recall from Chapter 5 (p. 160) that each treatment factor may have a number of levels associated with it. Each combination of levels is called a **treatment**. The set of available treatments determines which comparisons can be made. Comparisons between the treatments tell the experimenter how each treatment affects (or does not affect) the response.

7.4.4 Experimental units

The **experimental units** are the objects or subjects to which the treatments are applied. In Bennett et al.'s [9] spatial variability study, the experimental units are the panels of glass. In Rowe and Hanson's [33] shotgun experiment, the experimental units are the pieces of cardboard that are used as targets. In Abaz et al.'s [44] drinking container/beverage experiment, the experimental units are the different people who drank from each container. In Wong's [45] and Hicks et al.'s [56] glass-breaking experiments, the experimental units were the panes of glass used for each shot.

7.4.5 Structure in experimental units

In many situations the experimental units are not completely homogeneous. That is, the experimental units may have observable and measurable differences between them. These differences could affect the results of the experiment. There are two main strategies for coping with this which we will discuss shortly. These are **blocking** and **randomization**. In general, however, if we can take account of any systematic structure that exists in the experimental units, then our experiment will be more efficient, in terms of the number of experimental units needed, and the comparisons will be more accurate.

7.4.6 Assignment of treatments to experimental units

When we talk about the assignment of treatments to experimental units we are talking about assigning combinations of the levels of the treatment factors. All of the examples we have seen so far have factorial structure, meaning that every combination of every level of the treatment factors is used at least once. This may not be the case in more complicated experimental designs. The experimenter must decide which treatment will be assigned to which experimental unit. There are two key objectives in this process. Firstly, the experimenter should try and exploit any structure that exists between the experimental units. Doing this makes the comparisons more fair and as precise as possible. Secondly, the experimenter needs to ensure that the analysis will be valid from a statistical point of view. This essentially reduces to the appropriate use of randomization where needed.

7.5 The principles of experimental design

Fisher [57] proposed three important principles that should be considered when designing an experiment. These principles are replication, blocking, and randomization. Sadly, most of these are overlooked.

7.5.1 Replication

Replication simply means that each treatment should be used more than once in the experiment. At its heart statistics is about the study of variability. Without replication we cannot assess variability. If we cannot assess variability, then we have no way of putting our observations into perspective or context. That is, we cannot judge whether the differences that we have observed could be due to random variation alone. Replication allows us to estimate the inherent variability in the data. Unfortunately, replication is often overlooked. There are a few situations where we can ignore replication, but they primarily relate to **screening experiments**. Screening experiments are used when there are a large number of treatment factors that the experimenter may wish to investigate. To do this a low-quality experiment is carried out and the factors that have the largest effects on the response are retained for more detailed experiments. Standard questions statisticians get asked are, "How many replicates should I do?" or "How big a sample should I take?" The response, as facetious as it seems, is, "As many as possible." The reason for this response is that the more replication we have we understand about the variability of the process, and the more sure we can be about any observed differences. My rule-of-thumb is "at least three replicates per treatment." Please do not interpret this as "three is all we ever need."

7.5.2 Blocking

Blocking is the statistical term for exploiting any structure that exists in the experimental units. If possible, the experimental units should be divided into groups, which we call the **blocks**. The units in each block should be more similar to each other than to units from other blocks. For example, if we are measuring the concentration of a naturally occurring hormone, such as GHB, and we have experimental subjects who are both female and male, then it may be sensible to block on gender. We use blocking to take into account differences that we know exist before the experiment is performed. The advantage in blocking is that if two units in the same block receive the same treatment then they can be compared more precisely than if the treatments were assigned to units in different blocks. Blocking is also efficient. That is, it allows us to improve our chances of detecting a difference between treatments without having to increase the overall size of the experiment. Many people have inadvertently used blocking when performing a paired t-test. A common application of a paired t-test is when we have *before* and *after* measurements on experimental subjects. For example, an experiment we perform in class sometimes is to get students to measure their pulse while standing, and then again after running on the spot for 60 seconds. In this example, the students are the blocks.

7.5.3 Randomization

Randomization refers to the use of a random process to allocate treatments to experimental units. We use randomization firstly and foremostly to avoid unconscious or systematic biases in the experimental units. That is, when we use randomization, each treatment has the same chance of being allocated to a good or bad experimental unit. Secondly, the use of randomization justifies an assumption we made very heavily in Chapter 5, namely, that of independence. Randomization means that our estimates will behave as though they were based on independent observations. We use randomization to account for the differences that we cannot block for. It can often be a good idea to use randomization even when there is no treatment. For example, Bennett [37] randomized the order in which she took samples from each panel of glass to avoid any bias that might be due to inter-day variation in the measurement process.

7.6 The description and analysis of experiments

In the ensuing sections, we will discuss different families or classes of experimental designs. All experimental designs can be described in terms of

- block structure, a description of the structure (if any) present in the experimental units,

- treatment structure, a description of the treatments,

- and allocation of treatments to experimental units.

The statistical analysis of any design depends entirely on these components. There are many "named" designs. That is, classes or families of designs that have special names. It is impossible to cover them all and ultimately it would not be useful to do so. Instead, we will cover some common designs that describe the basic structures that are present in more complex variants.

7.7 Fixed and random effects

You may recall that equation (5.16) on page 208 unified the linear model. A simpler way of regarding *every* model is

$$data = signal + noise \tag{7.1}$$

In this chapter we model the signal as the expected value of the response (or data) given the treatment factors. The noise models the variability in the response as a function of the blocking factors. In the models we have seen thus far, we use **fixed effects** or terms to model the signal, and **random effects** to model the noise. There is often a distinction made between fixed effects models, random effects models, and mixed models. These distinctions are somewhat artificial. All models contain a mixture of fixed effects and random effects, and hence are **mixed models**. The distinction seems to be if the error structure for the model is very simple and there is no blocking (or each experimental unit is a block), then the model is regarded as a **fixed effects model**. If the model has a block structure but no treatment, then it can be regarded as a pure **random effects model**. This last distinction is sometimes called a **variance components model**. The difference, if any, between random effects and variance components models is that in random effects models there is the concession that you might be interested in the effects themselves, rather than just which part of the variability in the data they model. Finally, if we have both treatment (fixed) effects and block (random) effects, then we have a mixed model.

7.8 Completely randomized designs

A completely randomized design (CRD) is an experiment where the experimental units are considered to be homogeneous. That is, every experimental unit is regarded as being the same as every other experimental unit. One way of thinking about a completely randomized design is that there is no block structure. Another is to think that there are as many blocks as there are experimental units. The latter is more useful in that it gets us into the mind-set of thinking about block structure. Recall that in general block structure can help us make our experiments more efficient in terms of resource use.

7.8.1 Examples

1. In Rowe and Hanson's experiment, there 60 shots fired (ignoring those made for prediction testing); therefore, there are 60 experimental units (the targets) and 60 blocks.

2. In Wong's experiment there were (in theory) 72 shots fired so there are 72 experimental units and 72 blocks.

3. In Bennett's experiment there are 490 measurements made on one pane of glass, and so 490 blocks. Bennett's experiment is a little different, however, in that there is a physical block. The window-pane was cut up into 7×7 grid, so one might regard the panels as the blocks.

The design of experiments

We will concentrate again on Rowe and Hanson's shotgun range estimation experiment. We concentrate on this because it has features that allow us to describe various parts of a CRD. Rowe and Hanson [33] were interested in estimating the distance at which a shotgun had been fired from the area of the pellet scatter pattern. As we saw in Chapter 5, their aim was primarily an inverse prediction problem: "Given the pellet scatter area how far away do we think the shooter was?" However, we can treat this as an experiment where we are interested in the effect of range (and shotgun brand) on pellet scatter pattern area. When we analyzed this data previously we treated the explanatory variable **range** as continuous. We would be also justified in treating **range** as a factor with five levels (10, 20, 30, 40, and 50 ft).

In this experiment we are interested in estimating μ_i, the mean pattern area, and the overall variability σ.

7.8.1.1 Block structure

In a completely randomized design, every experimental unit is its own block. Therefore, the block structure in this experiment consists of

- one blocking factor: **target** with 60 levels corresponding to the blocking factor, one random variable ϵ with 60 observed values ($\epsilon_1, ..., \epsilon_{60}$).

- in the language of random effects, we assume that the blocks follow a normal distribution with mean zero, and variance σ^2, i.e., $\epsilon_i \sim N(0, \sigma^2)$.

7.8.1.2 Treatment structure

The treatment structure for this experiment is that the pellet pattern area is affected by the range at which the shotgun was fired. Therefore, the treatment structure consists of a single treatment factor **range** with five levels (10, 20, 30, 40, 50).

7.8.1.3 Randomization

In this experiment, the ranges should have been assigned to targets completely at random. That is, some random process such as a random number generator should have been used to divide the targets into groups of size 12, and then the first 12 shots could be taken at 10 ft, the second 12 shots at 20 ft, and so on. We have no way of knowing whether Rowe and Hanson actually did this short asking them.

If we were to perform this experiment ourselves, we could use the **sample** function in R to create a random permutation of the targets.

```
> target = 1:60
> perm = matrix(sample(target), nr = 12)
> perm
```

          [,1] [,2] [,3] [,4] [,5]

```
 [1,]  58  33  46  14  25
 [2,]  27  23  35  50  59
 [3,]  20  24  44  11  21
 [4,]  18  48  32   2  39
 [5,]  47   5  52   8  43
 [6,]   3  54  41  45  30
 [7,]  49  22   4  10  34
 [8,]  16  28  42  51  19
 [9,]  40  55  37  29  17
[10,]  60  57  15  31  12
[11,]  56  13  26  53  36
[12,]  38   7   1   6   9
```

The `matrix` command helps us divide the random permutation into groups. Each target in a column of the matrix above will be fired at from the same distance. That is, targets 58, 27, 20, ... are fired at from 10 ft, targets 33, 23, 24, ... are fired at from 20 ft and so on.

7.8.1.4 Analysis in R

We will not spend an excessive amount of time reanalyzing the data. However, in this chapter we want to explicitly account for block structure. The way to do this is to use the `aov` function instead of the `lm` function. The `aov` function has a special option `Error` which may be used in the formula to specify the block structure. This in turn will alter the analysis of variance table.

The data are stored in the data frame `shotgun.df`. We wish to use only the training part of the data so firstly we will extract those observations into a new data frame.

```
> data(shotgun.df)
> train.df = subset(shotgun.df, expt=="train")
```

The data frame does not have any information about the block structure so we need to create that. It also will treat `range` as a continuous variable unless we tell it otherwise. The command `as.factor` can be used to *coerce* `range` into a factor.

```
> train.df = within(train.df, {
    range = as.factor(range)
    target = factor(1:60)
})
```

We have specified the treatment and block structure so now we can fit the model using the `aov` command

```
> shotgun.aov = aov(log(sqrt.area) ~ range +
    Error(target), data = train.df)
```

The ANOVA table is obtained from R by either explicitly using the `summary` command, or simply typing the name of the fitted `aov` object. We will stick to the formal approach of using `summary`. Note that (rather oddly) `anova` will return an error message.

```
> summary(shotgun.aov)

Error: target
            Df  Sum Sq  Mean Sq  F value  Pr(>F)
range        4   16.21     4.05      115  <2e-16 ***
Residuals   55    1.94     0.04
---
Signif. codes:  0 '***' 0.001 '**' 0.01 '*' 0.05 '.' 0.1 ' ' 1
```

You will probably notice that this output looks very similar to that of the `anova` command. This is because everything we have done so far has been treated as a completely randomized design. The output becomes more complex with a more complex block structure. As before, the line for `range` summarizes the F-test for the difference between the mean pattern areas at different ranges. There is very strong evidence of a difference.

7.8.1.5 Factorial treatment structure

We actually have two factors in this experiment, `range` and `gun`. `gun` has two levels which represent the two brands of shotgun that were used. We also know that each gun was used at each of the firing ranges. This type of treatment structure, where every level of each factor is used with every level of the other factors, is called a **factorial design**. The treatment factors `gun` and `range` are said to be **crossed**. The treatment structure using both treatment factors is $R*G$, where R is `range` and G is `gun`. There are $5 \times 2 = 10$ treatments corresponding to all the possible unordered combinations of the levels of the treatment factors. One important consideration in experimental design is the question of whether we have sufficient data or replication to estimate all of the effects. Experimental design experts think about this problem in terms of degrees of freedom. In Chapter 5 we saw that $R * G$ expands into

$$R + G + R \times G$$

In these types of formulae, any term that consists of only one letter is called a main effect. R and G are the main effects in this model. Any term that consists of two or more letters is called an interaction. An additional qualifier is occasionally added to the interaction relating to the number of letters. In this model $R \times G$ is the two-way or 2-way interaction. The qualifier is more commonly added when there is more than one interaction term in the model. This may occur in an experiment involving three or more factors. The degrees of freedom associated with each of the terms are

1. $I_i - 1$ for each of the main effects where I_i is the number of levels associated with the ith factor $i = 1, 2, \ldots$.

2. the product of the degrees of freedom for each terms involved in the interaction.

So, in this model there are $5 - 1 = 4$ degrees of freedom associated with **range**, $2 - 1 = 1$ degrees of freedom associated with **gun**, and $4 \times 1 = 4$ degrees of freedom associated with the interaction of **range** and **gun**. There are $4 + 1 + 4 = 9$ degrees of freedom associated with the treatment structure.

The block structure also affects the degrees of freedom. However, we will worry about this when the block structure is more complex. In a CRD there is only one **error stratum**. The total degrees of freedom in this stratum relate to the total number of experimental units N, or n, and also to how many additional parameters we must estimate in order to estimate the parameters associated with the block structure. In a CRD we assume that $\epsilon_i \sim N(0, \sigma^2)$. That is, the block effects are normally distributed around 0 with a variance of σ^2. Therefore, the parameter we estimate that is associated with block structure is the variance. We know that in order to estimate the variance we need to estimate the grand mean. Therefore, there is a 1 degree of cost to estimate the grand mean and there are $N - 1$ degrees of freedom from which the remaining effects and parameters can be estimated.

The reason I have spent the time explaining degrees of freedom is because the residual degrees of freedom dictate how precise our comparisons are. If the residual degrees of freedom are large, then our comparisons will be more accurate.

```
> shotgun.aov1 = aov(log(sqrt.area) ~ range *
    gun + Error(target), data = train.df)
> summary(shotgun.aov1)

Error: target
           Df Sum Sq Mean Sq F value  Pr(>F)
range       4  16.21    4.05  116.62  <2e-16 ***
gun         1   0.02    0.02    0.62    0.43
range:gun   4   0.18    0.05    1.32    0.27
Residuals  50   1.74    0.03
---
Signif. codes:  0 '***' 0.001 '**' 0.01 '*' 0.05 '.' 0.1 ' ' 1
```

We have $N = 60$ experimental units in this experiment. One degree of freedom is used for the block structure, 9 degrees of freedom are used for the treatment structure, and hence (as we can see from the ANOVA table above) there are $50 = 60 - (4+1+4) - 1$ degrees of freedom available to estimate the residual standard error. If we only had three shots per treatment combination, then there would have been 20 degrees of freedom available to estimate the residual mean square. By increasing the number of replicates per treatment we have made the comparisons our experiment approximately 1.6 times more precise, because the estimated standard deviation will be 1.6 times smaller.

7.8.1.6 Interaction plots

Interaction plots are a useful way of exploring data in experiments with two treatment factor, and in particular the behavior of the interactions. The plot is constructed using the following steps:

1. The factor with the largest number of levels is labelled factor 1, and the other factor 2

2. Use the levels of factor 1 to construct the x-axis

3. Use the range of the response to construct the y-axis

4. Plot the average response level versus factor 2

5. Join the points that have the same level of factor 2

If there is no interaction between the two treatment factors, then the lines should be roughly parallel. I suggest that you use the command `interactionPlots` from the `s20x` library rather than the built in command `interaction.plot`. The former has a nice formula interface, meaning you can get a plot using the same formula (minus the `Error` term) you used to fit the model. Secondly, `interactionPlots` draws error bars around the means, which is useful for judging the significance of differences. We can see from Figure 7.1 that the lines for each different shotgun are roughly parallel. The lines do overlap for the shots fired at 50 ft. However, this is due entirely to the significant outlier for the Stevens gun. Removal of this data point removes all overlap. We can also see that each interval for the two guns overlaps for every level of the treatment factor `range` indicating that the factor `gun` is not significant. This is confirmed by the output we saw previously in the ANOVA table.

7.8.1.7 Quantitative factors

In many circumstances a treatment factor will have levels that can be interpreted quantitatively. For example, in the shotgun experiment the treatment factor `range` has a clear quantitative meaning. It would be useful to take advantage of this information in our modeling process. **Orthogonal polynomial contrasts** can help us detect whether there is any polynomial trend in our data. We are primarily interested in a linear trend, but it would be advantageous to know if there is any non-linearity that can be modeled by a polynomial. For example, we saw early in the regression modeling, that the shotgun data have a hint of a quadratic trend in it. In Chapter 5 (p. 178) we discussed linear contrasts, but we did not explicitly discuss orthogonality.

Suppose we have two contrasts

$$C_1 : a_1\mu_1 + a_2\mu_2 + \cdots + a_T\mu_T$$
$$C_2 : b_1\mu_1 + b_2\mu_2 + \cdots + b_T\mu_T$$

then C_1 and C_2 are **orthogonal** if

$$\sum_{i=1}^{T} a_i b_i = 0$$

For example, if $T = 4$, then the contrasts

C_1	1	−1	0	0
C_2	1	1	−2	0
C_3	1	1	1	−3

are mutually orthogonal. That is, every pair of contrasts is orthogonal to each other. Orthogonality is useful because it means estimates of the contrasts are independent and thus uncorrelated. Orthogonal polynomial contrasts can be used to test for polynomial trends in a quantitative factor. There is a set $T-1$ orthogonal polynomial contrasts for a treatment factor with T levels. These contrasts divide the variation associated with the treatment up into linear, quadratic, ... components. For example, in our shotgun data, the treatment

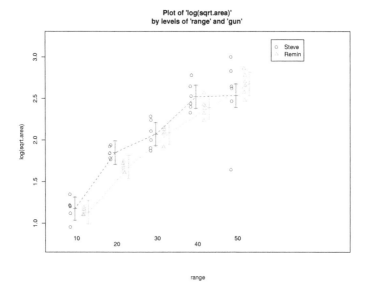

FIGURE 7.1: Interaction plot for the shotgun experiment

factor range has five levels; hence, there is a set of $5 - 1 = 4$ orthogonal polynomial contrasts that can be used to test for linear, quadratic, cubic, and quartic trends. There are infinitely many such sets of contrasts, but this is not something we need to worry about. R has a special command ordered which is used to specify that the levels of a factor have an ordering. It makes no difference to any of the computation we have done so far but is used to calculate the polynomial contrasts. We use the ordered command and a special option of summary estimate polynomial contrasts.

```
> train.df = within(train.df, {
    range = ordered(range)
})
> shotgun.aov = aov(log(sqrt.area) ~ range,
    data = train.df)
> summary(shotgun.aov, split = list(range = list(L = 1,
    Q = 2, C = 3, Q4 = 4)))

            Df Sum Sq Mean Sq F value   Pr(>F)
range        4  16.21    4.05  114.72 < 2e-16 ***
  range: L   1  15.51   15.51  438.89 < 2e-16 ***
  range: Q   1   0.63    0.63   17.81 9.2e-05 ***
  range: C   1   0.01    0.01    0.14    0.71
  range: Q4  1   0.07    0.07    2.04    0.16
Residuals   55   1.94    0.04
---
Signif. codes:  0 '***' 0.001 '**' 0.01 '*' 0.05 '.' 0.1 ' ' 1
```

Notice the special syntax for the summary command. This is actually an option for the summary.aov command that is called when the argument supplied to summary is an aov object. The labels I have supplied L, Q, etc., are just labels. They could be anything you like. For example, I could have used Linear, Quadratic, Cubic etc. or nothing split = list(range = list(1,2,3,4)). The labels simply make the output easier to read. The important part is the numbers 1,2,3,4. The range = list part of the code tells R that it is the treatment variability associated with range that we wish to decompose. This is the only choice for this model because range is the only treatment factor that can be treated as quantitative. However, do be aware that in other models there may be more choices, and that one can even decompose interactions. See help(summary.aov) for some example of this.

We can see from the output that the linear range: L and range: Q quadratic terms are significant indicating that there are linear and quadratic trends present in the data. However, we can also see that the treatment sum of squares linear trend is 16.2. Of this, the linear contrast takes 15.5 or 95.6%. The quadratic term takes 3.9%, on the other hand, so as before, we would probably be justified in ignoring the quadratic trend.

7.9 Randomized complete block designs

Randomized complete block designs (RCBDs) are an extremely popular choice for situations where the experimenter wishes to take (simple) block structure into account. The idea, as we discussed earlier, is to group the experimental units into blocks so that there is less variability among members of the same block. In doing this we increase the precision between the treatment comparisons.

7.9.1 Block structure

We need two blocking factors to describe the block structure for a randomized complete block design. We need a block factor which tells us which block each unit is in, and a unit factor to distinguish between units within a block. The block structure is described as units **nested** within blocks, and is written

$$blocks/units$$

The units within each block are assumed to be homogeneous. The $*$ symbol is the **crossing operator**. The $/$ symbol is the **nesting operator**. Just like multiplication and division $*$ is commutative and $/$ is not. Commutative means that $a*b = b*a$. One way to think about the nesting operator is that the units *divide* blocks.

7.9.2 Data model for RCBDs

The data model for a randomized complete block design is

$$y_{ij} = E[y_{ij}] + b_i + \epsilon_{ij}$$

where b_i is the random effect of block i and ϵ_{ij} is the random effect of unit j in block i. We assume that

$$b_i \sim N(0, \sigma_b^2)$$
$$\epsilon_{ij} \sim N(0, \sigma^2)$$

7.9.2.1 Example 7.1—Annealing of glass

Rushton [58] was interested in the effect of annealing on the refractive index of glass. It is well known that annealing (heating and cooling) of float glass has a substantial effect on the RI of the glass. Rushton cut a single pane of float glass into 150 squares. Three fragments were sampled from each square and the RI was determined pre- and post-annealing. We know from the work of Bennett et al. [9] that there is variability in RI over a pane of glass, although it is not systematic. Therefore, we wish to take this variability into account in our examination of the effect of annealing on RI.

7.9.2.2 Treatment structure

We want to allocate the treatments so that comparisons are made within blocks rather than between blocks. We achieve this by allocating each treatment to the same number of units (usually one) within each block. In this experiment there are several levels of blocking. Initially, we will work under the assumption that just one fragment has been sampled from each square so that each fragment is a block. You can now think each fragment as being broken into two units, one of which is annealed and one which is not. The choice of which unit is annealed is random. In this experiment, it is more likely that the fragment had its RI determined, then it was annealed and the RI was determined again. This is obviously not random. This falls under the category of **repeated measures**. Repeated measures models are used in situations where two or more measurements are made on each experimental unit. In simple circumstances, such as where there are only two treatments, the models and interpretation for both situations are the same. Therefore, our treatment structure is a single treatment factor `anneal` with two levels: `pre` and `post`

7.9.2.3 Block structure

As noted there are actually three levels of blocking in this experiment: squares, fragments within squares, and units within fragments. However, initially, we will just take one fragment from each square so that the fragment blocking factor is synonymous with the square blocking factor. Therefore, our (initial) blocking structure is

$$fragment/unit$$

7.9.2.4 Tutorial: Analysis in R

1. The data for this example are in the data frame `anneal.df`. Load the data.

   ```
   > data(anneal.df)
   > names(anneal.df)
   
   [1] "temp"    "ri"     "anneal"
   ```

2. The data frame has no blocking information but is stored so that there are 450 `pre` measurements and 450 `post`, with 3 measurements for each of the 150 squares. In our initial analysis, we only want one fragment per square; therefore, we will take a subset of the data. We will also scale the RI values for the same reason as before, numerical stability.

   ```
   > simple.df = anneal.df[3 * (0:299) + 1, ]
   > simple.df = within(simple.df, {
         ri = (ri - 1.518) * 1e+05
     })
   ```

3. Now we need to create our blocking factors `fragment` and `unit`. The data are organized so that the first unit for each fragment is in the first 150 observations, and the second in the second 150 observations. Each fragment is in order in each block.

```
> simple.df = within(simple.df, {
    fragment = factor(rep(1:150, 2))
    unit = factor(rep(1:2, rep(150, 2)))
})
```

4. Now we are ready to fit the model.

```
> anneal.fit = aov(ri ~ anneal + Error(fragment/unit),
    data = simple.df)
> summary(anneal.fit)

Error: fragment
          Df Sum Sq Mean Sq F value Pr(>F)
Residuals 149   5086    34.1

Error: fragment:unit
          Df Sum Sq Mean Sq F value Pr(>F)
anneal      1  99408   99408    3923 <2e-16 ***
Residuals 149   3776      25
---
Signif. codes:
    0 '***' 0.001 '**' 0.01 '*' 0.05 '.' 0.1 ' ' 1
```

5. You will notice that we now have two error strata: one for the fragments and one for the units within fragments. The residual mean square from the `fragment:unit` stratum estimates σ^2. The residual mean square from the `fragment` stratum estimates $\sigma^2 + n_u \sigma_b^2$ where n_u is the number of units per block. Therefore,

$$\widehat{\sigma}^2 = 25.3$$
$$\widehat{\sigma}^2 + 2\widehat{\sigma}_b^2 = 34.1$$
$$\widehat{\sigma}_b^2 = (34.1 - 25.3)/2$$
$$= 4.4$$

We can use these numbers. The variance of an experimental unit is the sum of the block variance and the individual variance $\sigma_b^2 + \sigma^2$. The covariance between units from the same block is σ_b^2.

6. As we noted the true block structure is slightly more complicated. We have units nested within fragments nested within squares. We write this as

$$squares/fragments/units$$

The design of experiments

There are 150 squares, three fragments per square, and two units per fragment, and so 150 × 3 × 2 = 900 observations. We can fit the model with this block structure just as easily as before. Note that the definition of fragment is different from the previous model.

```
> anneal.df = within(anneal.df, {
            ri = (ri-1.518)*100000
            square = factor(rep(rep(1:150,rep(3,150)),2));
            fragment = factor(rep(letters[1:3],150));
            unit = factor(rep(1:2, c(450,450)))
          })
> anneal.fit1 = aov(ri~anneal+Error(square/fragment/unit),
              data = anneal.df)
> summary(anneal.fit1)

Error: square
           Df Sum Sq Mean Sq F value Pr(>F)
Residuals 149   7593      51

Error: square:fragment
           Df Sum Sq Mean Sq F value Pr(>F)
Residuals 300   6542    21.8

Error: square:fragment:unit
           Df Sum Sq Mean Sq F value Pr(>F)
anneal      1 299096  299096   11502 <2e-16 ***
Residuals 449  11675      26
---
Signif. codes:
       0 '***' 0.001 '**' 0.01 '*' 0.05 '.' 0.1 ' ' 1
```

7. To compare the treatment means, or in this case the effect of annealing, we need to firstly estimate the means, which can be done with model.tables.

```
> model.tables(anneal.fit1, "means")

Tables of means
Grand mean

97.903

 anneal
anneal
   post     pre
 116.13   79.67
```

Secondly, we need to determine the standard error with which to compare the difference. We use the residual mean square from the square:fragment:unit stratum. The variance of the difference in two treatment means is $2\sigma^2/2 = \sigma^2$. Given that the (scaled) difference in the means is about 34, and the (scaled) standard deviation is 5.1, there is obviously a massive effect due to annealing.

7.9.2.5 Example 7.2—DNA left on drinking containers

You may recall that in the analysis of the experiment of Abaz et al. [44] in Chapter 5 we noted that it was incorrect to treat the subject variable, person as a simple treatment factor. Now that we have the skills we can correctly block on subjects.

1. Firstly, we need to load and manipulate the data set so that it has the form we need.

   ```
   > data(abaz.df)
   > sumQ = rowSums(abaz.df[, 11:21])
   > sumQ[sumQ == 0] = NA
   > abaz.df = data.frame(log.quantity = log10(sumQ),
         container = abaz.df$ab.sample, person = abaz.df$person,
         time = factor(abaz.df$time, levels = c(24,
             48)))
   > abaz.df = abaz.df[-which(abaz.df$container ==
         "R"), ]
   > abaz.df = within(abaz.df, {
         unit = factor(rep(rep(1:9, rep(2, 9)),
             6))
   })
   ```

2. Our blocking structure in this experiment is actually quite a bit more sophisticated than I initially let on. We have six test subjects who have DNA quantity measured at two different times (24 hours and 48 hours). Each subject is exposed to nine different beverages. The way that I think of this is that each person has two different time persons with nine "slots" or units for each beverage. Therefore, our blocking structure is

   $$person/time/units$$

   What makes this difficult is that this experiment has seven missing observations. This makes our experiment **unbalanced**, which makes the estimation process (slightly) harder. R will give us warning messages and it is unclear whether the results can be trusted. There are a variety of ways to deal with this, but R is not capable of most of them. In this circumstance, because the proportion of missing data are relatively low, we will replace the missing values with the average for the person.

```
> idx = which(is.na(abaz.df$log.quantity))
> for (i in idx) {
    p = abaz.df$person[i]
    i2 = which(abaz.df$person == p)
    abaz.df$log.quantity[i] = mean(abaz.df$log.quantity,
        na.rm = T)
}
```

3. We can now fit the model

```
> abaz.fit = aov(log.quantity ~ container +
    Error(person/time/unit), data = abaz.df)
> summary(abaz.fit)

Error: person
          Df Sum Sq Mean Sq F value Pr(>F)
Residuals  5   4.44   0.888

Error: person:time
          Df Sum Sq Mean Sq F value Pr(>F)
Residuals  6   0.81   0.135

Error: person:time:unit
          Df Sum Sq Mean Sq F value  Pr(>F)
container  8   5.81   0.727    5.45 1.4e-05 ***
Residuals 88  11.73   0.133
---
Signif. codes:
         0 '***' 0.001 '**' 0.01 '*' 0.05 '.' 0.1 ' ' 1
```

4. We can perform a crude analysis to see how much more efficient our block design is. To do this we fit a completely randomized design and compare the residual mean squares.

```
> abaz.crd.fit = aov(log.quantity ~ container,
    data = abaz.df)
> summary(abaz.crd.fit)

          Df Sum Sq Mean Sq F value  Pr(>F)
container  8   5.81   0.727    4.24 0.00021 ***
Residuals 99  16.99   0.172
---
Signif. codes:
         0 '***' 0.001 '**' 0.01 '*' 0.05 '.' 0.1 ' ' 1
```

The ratio of the residual mean squares is $0.172/0.133 = 1.29$. This says, roughly, that we would need about 30% more replicates in a completely randomised design to achieve the same precision as the RCBD.

7.9.2.6 Example 7.3—Blood alcohol determination

For example, Miller et al. [59] were interested in the effect of adding sodium fluoride (NaF) to blood samples that had been submitted for blood alcohol determination. There was speculation that this process of "salting" caused higher blood alcohol readings. Miller et al. performed three experiments, and the results of this example come from their Table 2. In Miller et al.'s second experiment, six volunteer subjects were given alcoholic beverages over a one-hour period in an attempt to raise their blood alcohol concentrations to approximately 0.08% to 0.10% W/V. Three tubes of blood were taken from each subject, and 0, 5, or 10 mg/mL of sodium fluoride were added to each tube. The blood alcohol concentration for each tube was determined twice.

7.9.2.7 Treatment structure

There is a single treatment factor in this experiment, NaF, which has three quantitative levels 0, 5, and 10 mg/mL. The treatments should be allocated randomly to each tube within subjects. As the treatment is quantitative with three levels, we can use orthogonal polynomial contrasts to look for linear and quadratic trends in the response.

7.9.2.8 Block structure

There are three blocking variables subject, tube, and rep. The replicate measurements from each tube are technical replicates. This means that although we can be specific about replication when we describe the blocking structure, the variation in the replicates will be included in the residual mean square for the tubes within subject stratum. We will discuss this issue in more detail when we perform the analysis. The replicates are nested within each tube and each tube is nested within each subject. Therefore, our blocking structure is

$$subject/tube/rep$$

although, as we will see

$$subject/tube$$

is actually sufficient.

7.9.2.9 Tutorial - analysis in R

(a) Load the data.

```
> data(salting1.df)
> names(salting1.df)
```

The design of experiments

```
[1] "subject" "tube"    "rep"     "conc"    "NaF"
```

(b) Numeric variables are not interpreted as factors by default in R, so we need to make sure our blocking and treatment factors are interpreted as such. In addition our treatment factor is quantitative, so we will use `ordered` rather than `factor`.

```
> salting1.df = within(salting1.df, {
    subject = factor(subject)
    tube = factor(tube)
    rep = factor(rep)
    NaF = ordered(NaF)
})
```

(c) Before we fit the model, we can look at an interaction plot of the data.

```
> library(s20x)
> interactionPlots(conc~subject*tube,
                   data = salting1.df)
```

The interaction plot is shown in Figure 7.2. We can see that there is no interaction between the blocking factors as the lines are almost parallel. It looks like it is definitely worthwhile blocking on subject as there is a large amount of variation between subjects but very little variation within.

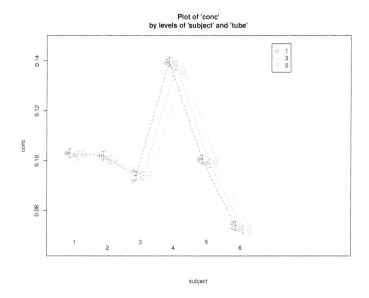

FIGURE 7.2: Interaction plot of salting experiment data

5. We can now fit our model.

```
> salt1.fit = aov(conc ~ NaF + Error(subject/tube/rep),
    data = salting1.df)
> summary(salt1.fit)

Error: subject
          Df Sum Sq Mean Sq F value Pr(>F)
Residuals  5 0.0132 0.00263

Error: subject:tube
          Df  Sum Sq  Mean Sq F value Pr(>F)
NaF        2 1.27e-05 6.33e-06    5.94   0.02 *
Residuals 10 1.07e-05 1.07e-06
---
Signif. codes:
        0 '***' 0.001 '**' 0.01 '*' 0.05 '.' 0.1 ' ' 1

Error: subject:tube:rep
          Df  Sum Sq Mean Sq F value Pr(>F)
Residuals 18 2.1e-05 1.17e-06
```

We can see from the output that the effect of the sodium fluoride treatment is tested in the `subject:tube` stratum. This is because each replicate, as we noted before, is a technical replicate. That it is a repeat measurement on a unit that has received the same treatment. The residual mean square for the `subject:tube:rep` stratum tells about the variation in the measurement process. The residual mean square from the subject:tube stratum is a combination of the variation in the measurement process, and the variation between tubes for the same subject. The residual mean square for the `subject` stratum is a combination of variation between subjects, variation between tubes, and measurement variation.

6. We can use the fact that `NaF` has a quantitative interpretation to look for polynomial trends in the data. `NaF` has two levels, so we can look for linear and quadratic trends using the `split` option with `summary`

```
> summary(salt1.fit, split = list(NaF = list(L = 1,
    Q = 2)))

Error: subject
          Df Sum Sq Mean Sq F value Pr(>F)
Residuals  5 0.0132 0.00263

Error: subject:tube
          Df  Sum Sq  Mean Sq F value Pr(>F)
NaF        2 1.27e-05 6.33e-06    5.94  0.020 *
```

```
NaF: L    1 1.07e-05 1.07e-05   10.00  0.010 *
NaF: Q    1 2.00e-06 2.00e-06    1.87  0.201
Residuals 10 1.07e-05 1.07e-06
---
Signif. codes:
          0 '***' 0.001 '**' 0.01 '*' 0.05 '.' 0.1 ' ' 1

Error: subject:tube:rep
          Df Sum Sq Mean Sq F value Pr(>F)
Residuals 18 2.1e-05 1.17e-06
```

We can see from the output that there is clear evidence of a linear trend with respect to increasing the amount of sodium fluoride. We could fit a regression model to estimate this effect, but in this case it will suffice to get the means.

7. We can get the means using `model.tables`

```
> model.tables(salt1.fit, "means")

Tables of means
Grand mean

0.10133

 NaF
NaF
      0       5      10
0.10217 0.10100 0.10083
```

Looking at the means, we can see there appears to be evidence of a decrease in the concentration measurements as the amount of sodium fluoride added increases.

8. If we re-draw the interaction plot, we can see what is driving this effect.

```
> interactionPlots(conc ~ NaF * subject, data = salting1.df)
> box()
```

We can see from the interaction plot in Figure 7.3 that really the whole downward trend is driven by subject number 4. The other subjects have little to no downward trend.

7.9.3 Randomized block designs and repeated measures experiments

Repeated measures describes experimental situations where two or more measurements are made on the same experimental unit. The statistical issue is that there may be a **carryover** effect between successive measurements on the same unit. That is, if subject 1 receives treatment 1 at time 0, the subject may still be affected by treatment 1 when they take treatment 2 at time 1. This is a major source of concern in clinical trials. A standard solution is to try and ensure that a sufficient amount of time has elapsed between measurements so that the effect of the previous treatment has **washed out** and is no longer an issue. This is equivalent to making sure that there is no dependence between successive measurements. If there are good reasons to think that the dependence between successive measurements is either very small or non-existent, then the experiment can be analyzed as a randomized complete block design where the measurements made on the same experimental unit are treated as different units within a block. If, however, this dependence cannot be ignored, then some attempt should be made to explicitly model it. We will look at an example where dependence is an issue.

7.9.3.1 Example 7.4—Musket shot

Jauhari et al. [60] performed a set of experiments that provided the motivation for the experiments performed by Rowe and Hanson [33]. In Jauhari et al.'s experiments, a sequence of 10 shots were fired from a 0.410-gauge musket. The cartridges used for each shot were loaded with 18 grains of cordite and 18 pellets of 0.46 mm in diameter. The shots were fired at a line of six targets that were equally spaced at 3 ft intervals, starting at 3 ft from the muzzle and ending at 18 ft. The response, labeled `size`, was calculated as the

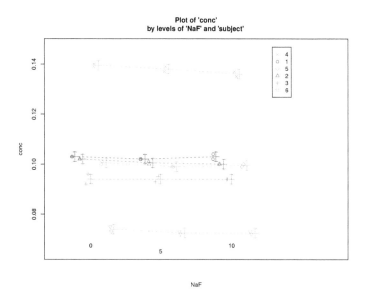

FIGURE 7.3: Interaction plot for salting data

average of the horizontal and vertical dispersion of the pellet pattern. In this experiment, we might think that there is variability from shot to shot. If there was one target for each shot, then we would not worry about this. However, there are six targets per shot. Secondly, there is spatial relationship between the targets. That is, each target is 3 ft apart. We might, again, reasonably hypothesize that the fact the pellets have passed through a target may have an effect on the velocity of the pellets and hence the pattern when they pass through the next target. The targets look like distinct experimental units, because there are after all 60 targets used. However, one might also think of them as a sequence of measurements in space, therefore, we should think about the correlation between successive targets from the same shot. Jauhari et al. [60] considered this issue and performed a pilot experiment. They fired a sequence of 10 shots through a set of six paper targets that were equally spaced from 3 ft to 18 ft and recorded the size of the pellet pattern at 18 ft. They then repeated the experiment with five of the targets removed and a single target at 18 ft. This data set is in the dafs package and is called jauhari1.df. A box plot of the data is shown in Figure 7.4. Jauhari et al. used a two-sample t-test to show that there is no difference due to the addition of the targets. A 95% confidence interval for the difference in the two means is $(-3.8, 1.0)$. This confidence interval clearly includes zero. This approach, however, does not model the explicit correlation that may exist between two targets. We can model first-order autocorrelation between successive targets

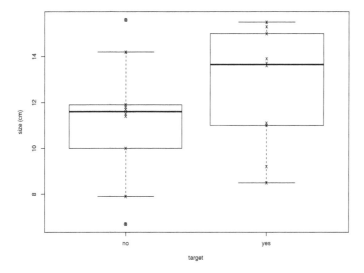

FIGURE 7.4: Pellet pattern size with and without intermediate targets for the Jauhari et al. data

by using a **lagged response** model. In such a model we add the (lagged) response to the model as an explanatory variable so that our model becomes

$$y_t = \rho y_{t-1} + \beta_0 + \beta_1 x_{1t} + \cdots + \beta_p x_{pt}, \ t = 2,..,n$$

The term ρ models the **first-order autocorrelation**. We say it is *first-order* because the current value y_t only depends on the value immediately before it y_{t-1}. We talk about **autocorrelation** because this describes correlation between values of the same random variable. If we modeled second-order autocorrelation, then the response would depend on y_{t-1} and y_{t-2}. This idea extends to third-order and so on. If we fit this model and find the parameter ρ is significant, then we have modeled the (first-=order) correlation structure. If ρ is not signficant, then really all this says is that there is no evidence of first-order linear dependence. There could be higher-order or non-linear dependence, but realistically a first-order linear dependence is going to be the most problematic. This approach works when the responses are equally spaced temporally or spatially. If this is not the case, then this approach may not be appropriate.

7.9.3.2 Treatment structure

There is only one treatment factor, `distance` with 6 levels (3 ft, 6 ft, ..., 18 ft).

7.9.3.3 Blocking structure

There are two blocking factors to deal with the different shots and the targets used within shots. Therefore, our blocking structure is

$$shot/target$$

The treatments are not applied randomly to the experimental units. It is reasonable to assume the targets are homogeneous.

7.9.3.4 Tutorial - Analysis in R

1. Load the data.

    ```
    > data(jauhari2.df)
    > names(jauhari2.df)
    ```

    ```
    [1] "shot"     "size"     "distance"
    ```

2. Plot the response `size` with respect to `distance`.

    ```
    > plot(size ~ distance, data = jauhari2.df)
    ```

    We can see from Figure 7.5 that the variation in the scatter pattern size increases with differences. Therefore, we will work with the logarithm of the distance. It is left to the reader to see that this equalizes the variances.

The design of experiments

3. We will initially do some regression modeling to deal with the potential of correlation between the targets from the same shot. In the first instance, we need to reorder the data so that success targets are in order for the same shot. If we do not do this, then fitting the lag response model becomes harder. We will also make shot a factor and create a factor called target.

```
> o = order(jauhari2.df$shot)
> jauhari2.df = jauhari2.df[o, ]
> jauhari2.df = within(jauhari2.df, {
      shot = factor(shot)
      target = factor(rep(1:6, 10))
  })
```

4. We want to set things up so that the lagged response contains the measurements for targets 1 to 5, and the response contains measurements for targets 2 to 6 so that our data model is

$$\log(size_{ij}) = \rho \log(size_{i,j-1}) + E[\log(size_{ij} \mid d_j)] + s_i + t_{ij}$$

where

- $i = 1, \ldots, 10, j = 2, \ldots, 6$
- s_i is the effect of the ith shot and $s_{ij} \sim N(0, \sigma_s^2)$

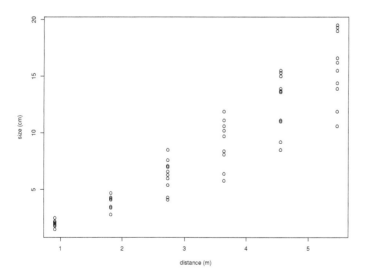

FIGURE 7.5: Scatter plot of pellet scatter pattern size by distance

- t_{ij} is the effect of the jth target in the ith shot and $t_{ij} \sim N(0, \sigma_t^2)$
- $E[\log(size_{ij})|d_j]$ is the fixed effect of firing the gun from a distance d_j. We will model this as quadratic in d_j, i.e., $\beta_0 + \beta_1 d_j + \beta_2 d_j^2$ as this seems to fit the behavior of the data and what the physics of the experiment would predict.

We store all of the indices of experiments where the target number is 1, and all of the experiment indices where the target number is 6.

```
> t1 = which(jauhari2.df$target == 1)
> t6 = which(jauhari2.df$target == 6)
```

Now we create a new data frame that has the appropriate structure.

```
> lag.size = jauhari2.df$size[-t6]
> j2.df = jauhari2.df[-t1, ]
> j2.df = within(j2.df, {
       lag.size = lag.size
  })
> j2.df[1:10, ]
```

	shot	size	distance	target	lag.size
11	1	3.4	1.82	2	1.5
21	1	5.4	2.73	3	3.4
31	1	8.1	3.64	4	5.4
41	1	11.0	4.55	5	8.1
51	1	13.9	5.46	6	11.0
12	2	2.8	1.82	2	2.5
22	2	4.1	2.73	3	2.8
32	2	6.4	3.64	4	4.1
42	2	9.2	4.55	5	6.4
52	2	11.9	5.46	6	9.2

You should be able to see from the output above that the first value in size is equal to the second value in lag.size, then second is equal to the third and so on.

5. Now we will fit the model. The lagged response should enter the model last. The reason for this is that we want to look for autocorrelation after we have removed any trend in the data.

```
> jauhari2.fit = lm(log(size) ~ shot + distance +
      I(distance^2) + log(lag.size), data = j2.df)
> anova(jauhari2.fit)

Analysis of Variance Table
```

```
Response: log(size)
               Df Sum Sq Mean Sq F value   Pr(>F)
shot            9   2.03    0.23  113.60  <2e-16 ***
distance        1  12.67   12.67 6387.73  <2e-16 ***
I(distance^2)   1   0.31    0.31  155.59   8e-15 ***
log(lag.size)   1   0.02    0.02    7.78  0.0083 **
Residuals      37   0.07    0.00
---
Signif. codes:
         0 '***' 0.001 '**' 0.01 '*' 0.05 '.' 0.1 ' ' 1
```

Note that we have, at this point, included shot as a fixed effect. While this is technically not what we want to do, it does adjust for the fact that there are 10 shots and not 60 shots. We can see from the ANOVA table that there is evidence of autocorrelation.

6. If we look at the coefficient for the lagged response,

```
> coef(jauhari2.fit)

  (Intercept)          shot2          shot3
     0.081200      -0.195825       0.126213
        shot4          shot5          shot6
     0.026398      -0.235153       0.304226
        shot7          shot8          shot9
     0.235556       0.157171       0.254977
       shot10       distance  I(distance^2)
     0.187145       0.651593      -0.046825
log(lag.size)
     0.161002
```

we can see there is about a correlation of about 0.16 after removing the distance effect between successive values. This is not especially high, but it would explain the (slightly) larger sizes that Jauhari et al. observed in their pilot experiment when the targets were in place.

This model is useless for predictive purposes as it stands because it the user needs to know the average pellet scatter pattern size if the gun had been fired 3 ft closer to the target.

We can overcome these short-comings by fitting a true random effects model with an **autoregressive** covariance structure. This requires the use of another R package.

7. **(Advanced)** This is the final analysis in the book. It probably goes far beyond the level of what we have encountered so far. However, I have included it to show the ability of statistics to model complex situations.

The model we will propose is similar to that proposed before.
$$\log(size_{ij}) = E[\log(size_{ij})|d_j] + s_i + t_{ij}$$
$$E[\log(size_{ij})|d_j] = \beta_0 + \beta_1 d_j + \beta_2 d_j^2$$
$$s_i \sim N(0, \sigma_s^2), i = 1, ..., 10$$
$$t_{ij} = \rho t_{i,j-1} + \nu_j$$
$$\nu_j \sim N(0, \sigma_\nu^2), j = 2, .., 6$$

To fit this model you will need to install and load the nlme package.

(a) Firstly, we will tell R how the data are grouped using the groupedData function.

```
> library(nlme)
> j3.df = groupedData(size ~ distance | shot,
      data = jauhari2.df)
```

The |shot says that shot is the grouping/blocking factor.

(b) Now we can fit the random effects, or mixed effects, if you are pedantic, model using the linear mixed effect model function lme.

```
> j.lme.fit = lme(log(size)~distance + I(distance^2),
    random = ~1|shot, data = j3.df,
    corr = corAR1(form = ~1|shot))
> j.lme.fit1 = lme(log(size)~distance + I(distance^2),
    random = ~1|shot, data = j3.df)
> j.lme.fit1 = lme(log(size)~distance + I(distance^2),
    data = j3.df)
```

We have fitted two models here, one with correlation structure and one without. I will briefly explain the parts of the call.

- distance + I(distance^2) is the fixed part of the model where we fit $E[\log(size_{ij})|d_j] = \beta_0 + \beta_1 d_j + \beta_2 d_j^2$.
- random = ~1|shot is the random part of the model that says the intercept β_0 varies randomly for each shot.
- corr = corAR1(form = ~1|shot) specifies the correlation structure. The corAR1 says we want a first-order autoregressive (AR(1)) correlation structure.

Therefore, the first model includes the AR(1) correlation and the second does not.

(c) We can compare the models using the ANOVA function

```
> anova(j.lme.fit, j.lme.fit1)
```

	Model	df	AIC	BIC	logLik	Test
j.lme.fit	1	6	-58.126	-45.868	35.063	

The design of experiments 289

```
j.lme.fit1     2 10 -81.909 -61.479 50.955 1 vs 2
          L.Ratio p-value
j.lme.fit
j.lme.fit1  31.784  <.0001
```

The significant *P*-value says that modeling the correlation is important.

(d) We can now get predicted values for each distance. We need to specify the shot number in order to use the `predict` function. This makes the prediction specific to the shot. To overcome this, we can make a prediction for each shot and then average over all of the shots. The code below finds the predicted sizes using four different models

   i. Linear regression with no account of shot variation
   ii. Linear regression with a fixed shot effect
   iii. Mixed effects/variance components model blocking on shot
   iv. Mixed effects model blocking on shot with an autoregressive correlation structure

```
> d = unique(jauhari2.df$distance)
> mod1 = lm(log(size) ~ distance + I(distance^2),
    data = jauhari2.df)
> mod2 = lm(log(size) ~ distance + I(distance^2) +
    shot, data = jauhari2.df)
> mod3 = aov(log(size) ~ ordered(distance) +
    Error(shot/target), data = jauhari2.df)
> p1 = exp(predict(mod1, data.frame(distance = d)))
> p3 = exp(model.tables(mod3, "means")$tables$"ordered
                (distance)")
> p2 = p4 = matrix(0, nc = 6, nr = 10)
> for (i in 1:6) {
    p2[, i] = exp(mean(predict(mod1, data.frame(distance
                    = d[i], shot = 1:10))))
    p4[, i] = exp(mean(predict(j.lme.fit,
        data.frame(distance = d[i], shot = 1:10))))
  }
> p2 = colMeans(p2)
> p4 = colMeans(p4)
> results = round(rbind(p1, p2, p3, p4), 3)
> colnames(results) = paste(seq(3, 18, by = 3),
    "ft", sep = "")
> rownames(results) = paste("Model ", 1:4, sep = "")
> results
           3ft    6ft    9ft   12ft   15ft   18ft
Model 1  2.013  3.678  6.093  9.148 12.449 15.355
```

```
Model 2 2.013 3.678 6.093 9.148 12.449 15.355
Model 3 2.004 3.697 6.141 9.056 12.429 15.405
Model 4 2.005 3.671 6.091 9.158 12.476 15.400
```

You might be forgiven for wondering if it was worth the effort.

7.10 Designs with fewer experimental units

The cost of carrying out experiments is often a major limiting step. Cost and resource constraints often lead researchers to compromise on the number of treatment factors and on the total number of experimental units. However, these compromises do not necessarily need to be made. In this section I will briefly discuss two types of design that can be used to reduce the number of experimental units: balanced incomplete block designs and 2^k designs. Of course, the decision to reduce the number of experimental units is not cost free. We will lose some precision either in the estimates or in the levels of the factors. However, these losses may be offset by the additional information gained.

7.10.1 Balanced incomplete block designs

Balanced incomplete block designs, or BIBDs, are useful when the number of units per block is less than the number of treatments. This might occur when we cannot afford to assign every treatment to every block. The BIBD compromise is to use a subset of the treatments in each block. The treatments are assigned to each block so that

- each treatment is replicated an equal number of times across all blocks
- and each pair of treatments occur together in the same block equally often

If treatments are assigned in this way, then this ensures that we can compare any pair of treatments with the same precision. The trade-off is a loss in efficiency. This is measured in the following way. If we let

- t be the number of treatments
- r be the number of replicates
- b be the number of blocks
- k be the number of units per blocks

Firstly, we note that
$$t \times r = b \times k$$

This identity is useful in weighing up the balance between our resources (blocks and units per blocks) and our costs (treatments and replicates).

The number of times each pair of treatments occur together in the same block is given by

$$\lambda = \frac{r(k-1)}{t-1}$$

and the **efficiency** of the design is defined as

$$e = \frac{t\lambda}{rk}$$

where $0 < e < 1$. The efficiency can be interpreted as comparing the performance of a BIBD to a comparable RCBD.

7.10.1.1 Example 7.5—DNA left on drinking containers II

In the experiment of Abaz et al., there were nine treatments and six subjects. Each subject received every treatment, and the results were measured at two different time periods resulting in 108 experiments. If we eliminated the time effect and reduced the number of treatments to six, then we can reduce the number of experimental units needed for a RCBD to 36. If we only give five of the treatments to each person, we only need 30 experimental units. To do this we would use a BIDB design with $t = 6$, $b = 6$, $k = 5$, and $r = 5$. This means each treatment is used five times. Each pair of treatments occurs together in four blocks as

$$\lambda = 5(5-1)/(6-1) = 4$$

The efficiency of this design is

$$e = (6 \times 4)/(5 \times 5) = 0.96$$

This means that the design is roughly 96% as efficient as a RCBD with five replicates of each treatment.

7.10.2 2^p factorial experiments

The factorial nature of most treatment structures makes testing many different factors impossible. Wong's experiment [45] is a good example. There were four levels for velocity, three levels for hardness and two levels for projectile shape therefore there were $4 \times 3 \times 2 = 24$ treatments. This means 24 experiments (or runs) were necessary for just one replicate of each treatment. I suspect that this experience is not uncommon among readers who have carried out experiments. There are several possible remedies:

1. Reduce the number of treatment factors;
2. Reduce the number of levels per treatment factor;

3. Do not replicate;

4. Only use some of the treatments.

We would like to avoid option 1 if possible. Option 3 is possible. The compromise for lack of replication is the inability to estimate interactions. If you have good prior knowledge that the interactions are negligible or non-existent, then this may be a viable option. However, in general, we do not know this is the case. Option 4 can be carried out using a class of designs called **fractional factorial designs**. In such designs the effects of associated with two or more treatments are **aliased** with each other. This means that we need to resort to special techniques to decide which of the aliased treatments are affecting the response. We will not discuss these designs further, although they are a subset of the designs we discuss in this section. The family of designs described by option 2 are called the 2^p family of designs or 2^p **factorial designs**. These designs are useful in situations where there are

- many treatment factors

- the variability in individual measurements is low

- observations are expensive to obtain

- observations are gathered in time order and often not simultaneously

We finish this section here without an examples or analysis because I have not found any in the literature, although I have seen a number of papers and conference talks that could definitely benefit from this sort of design.

As I noted at the start of this chapter, there are many different types of experimental designs, each of which can deal with particular experimental situations. If your own experiment is not covered by the designs that I have mentioned here then I suggest you pick up a basic book on experimental design.

7.11 Further reading

There are many books on experimental design. Unfortunately, as I previously noted, far too many of them concentrate on the underlying linear algebra. While this important, it is often not the primary concern of the practicing scientist. Two books that take a more rigorous approach than this chapter while still being (a) reasonably applied and (b) relatively up to date are *Design and Analysis of Experiments* by Montgomery [61] and *Introduction to Design and Analysis of Experiments* by Cobb [62]. Mixed and random effects models are a very important topic, and one that could be covered in a great deal more detail. Faraway [46] has a few pages on the R package lmer, which is probably more up to date than nlme. Zuur et al. [63] have a relatively new book *Mixed Effects Models and Extensions in Ecology with R* that has a lot of code, examples, and practical advice.

Bibliography

[1] G. Gustafson. Age determinations on teeth. *Journal of the American Dental Association*, 41(1):45–54, 1950.

[2] J. S. Buckleton, C. M. Triggs, and S. J. Walsh. *Forensic DNA Evidence Interpretation*. CRC Press, Boca Raton, FL, 2005.

[3] I. W. Evett and B. S. Weir. *Interpreting DNA Evidence: Statistical Genetics for Forensic Scientists*. Sinauer Associates, Sunderland, MA, 1998.

[4] C. G. G. Aitken and F. Taroni. *Statistics and the Evaluation of Evidence for Forensics Scientists*. Statistics in practice. Wiley, Chichester, 2nd edition, 2004.

[5] F. Leisch. Sweave: Dynamic generation of statistical reports using literate data analysis. In Wolfgang Härdle and Bernd Rönz, editors, *Compstat 2002—Proceedings in Computational Statistics*, pages 575–580. Physica Verlag, Heidelberg, 2002.

[6] A. S. C. Ehrenberg. *Data Reduction: Analysing and Interpreting Statistical Data*. Wiley, London; New York, NY, 1975.

[7] F. J. Anscombe. Graphs in statistical analysis. *The American Statistician*, 27(1):17–21, 1973.

[8] D. P. Kalm. The minimum daily adult, 2008. [Online; accessed 01-January-2010].

[9] R. L. Bennett, N. D. Kim, J. M. Curran, S. A. Coulson, and A. W. N. Newton. Spatial variation of refractive index in a pane of float glass. *Science & Justice*, 43(2):71–76, April 2003.

[10] R. Palmer and S. Oliver. The population of coloured fibres in human head hair. *Science & Justice*, 44(2):83–88, April 2004.

[11] W. S. Cleveland and R. McGill. Graphical perception: Theory, experimentation, and application to the development of graphical methods. *Journal of the American Statistical Association*, 79(3):531–554, 1984.

[12] A. W. N. Newton, J. M. Curran, C. M. Triggs, and J. S. Buckleton. The consequences of potentially differing distributions of the refractive indices of glass fragments from control and recovered sources. *Forensic Science International*, 140(2–3):185–193, March 2004.

[13] B. Found and D. K. Rogers. Investigating forensic document examiners' skill relating to opinions on photocopied signatures. *Science & Justice*, 45(4):199–206, 2005.

[14] P. Murrell. *R Graphics*. Chapman & Hall, Boca Raton, FL, 1st edition, July 2005.

[15] D. Sarkar. *Lattice: Multivariate Data Visualization with R*. Springer, 1st edition, August 2008.

[16] C. J. Wild and G. A. F. Seber. *Chance Encounters: A First Course in Data Analysis and Inference*. John Wiley, New York, NY, 2000.

[17] D. S. Moore and G. P. McCabe. *Introduction to the Practice of Statistics*. W. H. Freeman, New York, NY, 5th edition, 2006.

[18] R. R. Sokal and F. J. Rohlf. *Biometry: The Principles and Practice of Statistics in Biological Research*. Freeman, New York, NY, 3rd edition, 1995.

[19] Wikipedia. Rai stones, 2009. [Online; accessed 27-December-2009].

[20] Student. The probable error of a mean. *Biometrika*, 6(1):1–25, 1908.

[21] Wikipedia. Poisson distribution, 2009. [Online; accessed 27-December-2009].

[22] R. A. Fisher. *Statistical Methods for Research Workers*. Oliver and Boyd, Edinburgh, 14th edition, 1970.

[23] J. M. Curran, T. N. Hicks, and J. S. Buckleton. *Forensic Interpretation of Glass Evidence*. CRC Press, Boca Raton, FL, 2000.

[24] S. M. Weinberg, D. A. Putz, M. P. Mooney, and M. I. Siegel. Evaluation of non-metric variation in the crania of black and white perinates. *Forensic Science International*, 151(2–3):177–185, July 2005.

[25] K. Törő, M. Hubay, and É. Keller. Extramedullary haematopoiesis in liver of sudden infant death cases. *Forensic Science International*, 170(1):15–19, July 2007.

[26] F. Yates. Contingency tables involving small numbers and the χ^2-test. *Supplement to the Journal of the Royal Statistical Society*, 1(2):217–235, 1934.

[27] R. A. Fisher. On the interpretation of χ^2 from contingency tables, and the calculation of p. *Journal of the Royal Statistical Society*, 85(1):87–94, 1922.

[28] C. G. G. Aitken, T. Connolly, A. Gammerman, G. Zhang, D. Bailey, R. Gordon, and R. Oldfield. Statistical modelling in specific case analysis. *Science & Justice*, 36(4):245–255, October 1996.

[29] G. E. P. Box and N. R. Draper. *Empirical Model-Building and Response Surfaces.* Wiley, 1987.

[30] S. Martin-de las Heras, A. Valenzuela, and E. Villanueva. Deoxypyridinoline crosslinks in human dentin and estimation of age. *International Journal of Legal Medicine*, 112(4):222–226, June 1999.

[31] R. G. Gullberg. Employing components-of-variance to evaluate forensic breath test instruments. *Science & Justice*, 48(1):2–7, March 2008.

[32] J. G. Eisenhauer. Regression through the origin. *Teaching Statistics*, 25(3):76–80, 2003.

[33] W. F. Rowe and S. R. Hanson. Range-of-fire estimates from regression analysis applied to the spreads of shotgun pellet patterns: Results of a blind study. *Forensic Science International*, 28(3-4):239–250, August 1985.

[34] P. J. Brown. *Measurement, Regression, and Calibration.* Number 12 in Oxford statistical science series. Clarendon Press, Oxford, England, 1993.

[35] N. Kannan, J. P. Keating, and R. L. Mason. A comparison of classical and inverse estimators in the calibration problem. *Communications in Statistics—Theory and Methods*, 36(1):83, 2007.

[36] S. M. Ojena and P. R. De Forest. Precise refractive index determination by the immersion method, using phase contrast microscopy and the Mettler hot stage. *Journal of the Forensic Science Society*, 12(1):315–329, 1972.

[37] R. L. Bennett. Aspects of the analysis and interpretation of glass trace evidence. Master's thesis, Department of Chemistry, University of Waikato, 2002.

[38] G. N. Wilkinson and C. E. Rogers. Symbolic description of factorial models for analysis of variance. *Applied Statistics*, 22:392–399, 1973.

[39] F. Mari, L. Politi, C. Trignano, M. Grazia Di Milia, M. Di Padua, and E. Bertoli. What constitutes a normal ante-mortem urine GHB concentration? *Journal of Forensic and Legal Medicine*, 16(3):148–151, April 2009.

[40] R. G. Miller. *Beyond ANOVA, Basics of Applied Statistics*. Wiley, New York, NY, 1986.

[41] G. S. James. The comparison of several groups of observations when the ratios of the population variances are unknown. *Biometrika*, 38(3/4):324–329, December 1951.

[42] B. L. Welch. On the comparison of several mean values: An alternative approach. *Biometrika*, 38(3/4):330–336, December 1951.

[43] R. J. Carroll and D. Ruppert. Prediction and tolerance intervals with transformation and/or weighting. *Technometrics*, 33(2):197–210, May 1991.

[44] J. Abaz, S. J. Walsh, J. M. Curran, D. S. Moss, J. Cullen, J. Bright, G. A. Crowe, S. L. Cockerton, and T. E. B. Power. Comparison of the variables affecting the recovery of DNA from common drinking containers. *Forensic Science International*, 126(3):233–240, May 2002.

[45] S. C. K. Wong. The effects of projectile properties on glass backscatter: A statistical analysis. Master's thesis, Department of Chemistry, University of Auckland, 2007. Forensic Science.

[46] J. J. Faraway. *Extending the Linear Model with R: Generalized Linear, Mixed Effects and Nonparametric Regression Models*. Chapman & Hall/CRC, Boca Raton, FL, 2006.

[47] J. K. Lindsey. On the use of corrections for overdispersion. *Journal of the Royal Statistical Society. Series C (Applied Statistics)*, 48(4):553–561, 1999.

[48] W. N. Venables and B. D. Ripley. *Modern Applied Statistics with S*. Springer, New York, NY, 4th edition, 2002.

[49] R. Kent and J. Patrie. Chest deflection tolerance to blunt anterior loading is sensitive to age but not load distribution. *Forensic Science International*, 149(2–3):121–128, May 2005.

[50] J. J. Faraway. *Linear Models with R*. Chapman & Hall/CRC, Boca Raton, FL, 2004.

[51] A. Berkowicz, S. Wallerstedt, K. Wall, and H. Denison. Analysis of carbohydrate-deficient transferrin (CDT) in vitreous humour as a forensic tool for detection of alcohol misuse. *Forensic Science International*, 137(2-3):119–124, November 2003.

[52] B. Levine, D. Green-Johnson, K. A. Moore, D. Fowler, and A. Jenkins. Assessment of the acuteness of heroin deaths from the analysis of multiple blood specimens. *Science & Justice*, 42(1):17–20, 2002.

[53] D. Firth. Bias reduction of maximum likelihood estimates. *Biometrika*, 80(1):27–38, 1993.

[54] Ioannis Kosmidis. *brglm: Bias reduction in binary-response GLMs*, 2007. R package version 0.5-4.

[55] R. A. Fisher. The arrangement of field experiments. *Journal of the Ministry of Agriculture*, 33:511, 1926.

[56] T. Hicks, F. Schütz, J. M. Curran, and C. M. Triggs. A model for estimating the number of glass fragments transferred when breaking a pane: experiments with firearms and hammer. *Science & Justice*, 45(2):65–74, April 2005.

[57] R. A. Fisher. *The Design of Experiments*. Oliver & Boyd, Edinburgh, 8th edition, 1966.

[58] K. P. Rushton. Analysis of the variation of glass refractive index with respect to annealing. Master's thesis, Department of Chemistry, University of Auckland, 2009. Forensic Science.

[59] B. A. Miller, S. M. Day, T. E. Vasquez, and F. M. Evans. Absence of salting out effects in forensic blood alcohol determination at various concentrations of sodium fluoride using semi-automated headspace gas chromatography. *Science & Justice*, 44(2):73–76, April 2004.

[60] M. Jauhari, S. M. Chatterjee, and P. K. Ghosh. Statistical treatment of pellet dispersion data for estimating range of firing. *Journal of Forensic Sciences*, 17(1):141–149, 1972.

[61] D. C. Montgomery. *Design and Analysis of Experiments*. Wiley, New York, NY, 7th edition, July 2008.

[62] G. W. Cobb. *Introduction to Design and Analysis of Experiments*. Wiley, New York, NY, June 2008.

[63] A. F. Zuur, E. N. Ieno, N. J. Walker, A. A. Saveliev, and G. M. Smith. *Mixed Effects Models and Extensions in Ecology with R*. Springer, New York, NY, 1st edition, March 2009.

Index

additive model, 199
adjusted R^2, 138
aliased, 292
alternative hypothesis, 89
analysis of covariance, 160
analysis of deviance, 255
    anova, 215
analysis of deviance table, 214
anchoring, 25
ANCOVA, 160
ANOVA
    one-way, 168
ANOVA identity, 208
association, 259
autocorrelation, 284
autoregressive, 287

back transform, 201
backward elimination, 144
balance of probabilities, 248
balanced, 182
balanced incomplete block designs, 290
bandwidth, 48
Bernoulli trial, 234
beta-hat, 120
bitmap, 74
block, 197
blocking, 261, 262
Bonferroni's correction, 174

calibration, 151
calibration line, 151
carryover, 282
categorical, 8
    nominal, 9

ordinal, 9
causation, 258, 259
cdf, 87
Central Limit Theorem, 82, 98, 101
χ^2-test of independence, 102
chi-squared test of independence, 102
classification, 239
classification rule, 239
coerce, 266
complementary log-log, 235
complete separation, 243
confidence intervals, 93
confusion matrix, 239
consistent, 218
contingency tables, 237
contrast
    linear, 269
    orthogonal, 269
    polynomial, 269
convergence, 182
CRAN, 27
    Comprehensive R Archive Network, 27
cross-classification, 14
crossed, 103, 267
crossing operator, 272
cross-tabulation, 14
cumulative distribution function, 87
cumulative frequency curve, 51

data model, 118
data set, 8
data structure, 39
data type, 38
    logical, 39

numeric, 39
string literal, 39
degrees of freedom, 85, 267
derived variables, 146
descriptive statistics
   correlation coefficient, 23
   interquantile distance, 20
   interquartile range, 13, 18
   IQR, 18
   lower quartile, 13
   maximum, 12
   mean absolute deviation, 12
   midspread, 13
   minimum, 12
   mode, 17
   quantile, 13
      empirical, 13
      q_α, 13
      sample, 13
   range, 18
   sample mean, 10
   sample median, 12
   sample standard deviation, 10
   sample variance, 12
   upper quartile, 13
design
   2^p, 292
   balanced incomplete block, 290
   completely randomized, 264
   fractional factorial, 292
   randomized complete block, 272
dichotomous, 8
different intercepts, 161
different slopes, 162
dispersion parameter, 219
distributional shape, 20
distributions
   F, 87
   χ^2, 87
   Bernoulli, 234
   binomial, 86
   gamma, 219
   Gaussian, 82
   negative binomial, 220
   normal, 82

   Poisson, 87
   quasi-Poisson, 220
   standard normal, 83
   Studentized range, 176
   Student's t, 85
dummy variables, 160

effects model, 168
efficiency, 291
empirical rule, 84
error, 118
error stratum, 268
error sum of squares, 209
escape, 39
experiment, 258
experimental units, 260

factor, 160
factor level, 160
factorial design, 267
factors, 260
"fall-off-the-cliff" effect, 109
first-order correlation, 284
Fisher's exact test, 108
Fisher's (protected) LSD, 174
fitted model, 118
fitted values, 118, 121
fitting a model, 119
five number summary, 119
fixed effect, 197
fixed effects, 264
fixed effects model, 264
fractional factorial designs, 292
full factorial design, 213
funnel shape, 197

generalized linear model, 211
GLM, 211
   binomial, 234
   negative binomial, 220
   Poisson, 213
   quasi-Poisson, 221
goodness-of-fit, 87, 103

half-normal plot, 218
homogeneity of variance, 134

Index

hypothesis testing, 88

iid, 123
independence, 87
independence rule, 103
independent, 94
indicator, 160
indicator function, 255
influence, 25, 126
interaction, 162
interaction effects, 196
interaction plot, 194
interaction terms, 196
intercept, 118
inverse regression, 151
IRLS, 254
iteratively reweighted least squares, 254

kernel density estimate, 47
    bandwidth, 48

lagged response, 284
large sample approximation, 107
level, 160
leverage, 25, 166
likelihood, 110, 253
likelihood ratio, 109, 110
linear contrast, 178
linear scale, 51
link function, 213
    canonical, 213
locally weighted regression, 61
loess, 61
log-likelihood, 253
log transformation, 197
logarithmic scale, 51
logistic function, 235
logistic regression, 86, 211, 234
logit, 235
lowess, 61

main effects, 196
match temperature, 153
maximal model, 254
maximum likelihood estimates, 253

maximum likelihood estimation, 253
means model, 168
mean squared prediction error, 146
mixed models, 264
MLE, 253
mode, 17
multicollinearity, 144
multiple comparisons, 171
multiple correlation coefficient, 23
multiple linear regression, 133, 137
multiplicative model, 199

near separation, 243
negative binomial regression, 211
nested, 204, 272
nesting operator, 272
normally distributed, 82
normal Q-Q plot, 127
normal quantile-quantile plot, 127
null hypothesis, 89
null model, 255
numerical instability, 171

object, 39
observation, 8
observational study, 258
ODBC, 32
odds, 81
odds ratio, 109
    confidence interval, 110
OR, 109
ogive, 51
OLS, 180
one-way ANOVA, 168
Open Database Connectivity, 32
ordinary least squares, 180
orthogonal, 270
orthogonal polynomial contrasts, 269
outlier, 21
over-dispersed, 216
over-parameterized, 138

2^p factorial designs, 292
parameter, 88
Pareto chart, 52
pdf, 87

percentage of deviance, 220
plots
    bar plot, 47, 51
        stacked, 56
    box and whisker plot, 50
    box plot, 50
        outside points, 50
        whiskers, 50
    histogram, 47
        breaks, 47
        class intervals, 47
    kernel density estimate, 47
        bandwidth, 48
        Gaussian kernel, 48
        KDE, 47
    pairs plot, 25
    pie graph
        divided, 55
        exploded, 55
        perspective, 55
    scatter plot, 51
plotting device, 74
Poisson regression, 211
polychotomous, 8
polynomial regression, 137
polynomial term, 137
power, 124, 177
predicted values, 121
preponderance of the evidence, 248
principal components analysis, 146
probability, 81
probability density function, 87
probability function, 87
probit, 235
P-value, 89
pseudo-F statistics, 221

quadratic model, 135
quadratic term, 137
qualitative, 8
    nominal, 9
    ordinal, 9
quantitative, 8
    continuous, 9
    discrete, 9

R
', 39
:, 40
?, 33
["] operator for lists, 36
[[ ]], 43
[] for vectors, 36
$, 43
$ operator, 36
abline, 71
    h =, 71
    v =, 71
anova, 141
aov, 199, 204, 266
approx, 190
arrows, 171
as.factor, 165, 174
    levels, 174
barplot, 65
box, 65
boxplot, 70
c, 34
chisq.test, 104
choose.dir, 30
coef, 167
colors, 67
cumsum, 65
data.frame, 39
density, 68
deviance, 219
diff, 35
edit, 31
exp, 148
F, 39
FALSE, 39
file.choose, 30
fisher.test, 108
fitted, 129
formula, 71
function, 38
glm, 174, 214
    family, 214
    family = binomial, 236, 238
    family = gaussian, 233
    family = poisson, 214

Index

family = quasipoisson, 221
glm.nb, 221
graphics device
   bmp, 76
   jpeg, 76
   pdf, 76
   png, 76
   postscript, 76
   tiff, 76
   win.metafile, 76
   win.print, 76
   windows, 76
graphics.off(), 76
halfnorm, 228
help.search, 34
hist, 68
I, 137
ifelse, 248
install.packages, 32
interaction.plot, 269
interactionPlots, 269
IQR, 35
is.na, 31
legend, 67
LETTERS, 41, 187
letters, 187
library, 32, 33, 150
lines, 65
list, 39, 43
lm, 72, 119
   subs, 185
loess, 72
lowess, 72
matrix, 39, 42
mean, 35
median, 35
model.tables, 199, 204, 281
month.abb, 187
month.names, 187
multipleComp, 184
NA, 31
names, 31
nlme
   gls, 288
   groupedData, 288

normcheck, 130
order, 72
ordered, 271
package
   brglm, 250
   dafs, 32
   ggplot2, 77
   Hmisc, 251
   lattice, 77
   lmer, 293
   MASS, 221
   nlme, 288
   RODBC, 32
   s20x, 149
pairs20x, 150
plot, 68, 71
   col=, 67
   lty, 70
   pch, 70
   xlim, 69
   ylim, 69
points, 70
ppoints, 130, 228
predict, 158
   se = TRUE, 188
qnorm, 130
qqline, 130
qqnorm, 129
qtukey, 177
quantile, 35
read.csv, 29
read.table, 32
rep, 40
residuals, 127, 129
rowSums, 66
sapply, 37
scan, 31
sd, 35
seq, 40
setwd, 30
sort, 36
split, 37
subset, 231
sum, 31
summary, 167

split, 271
summary.aov, 271
T, 39
t.test, 96
    conf.level, 175
    var.equal, 175
table, 66
text, 65
TRUE, 39
unlist, 68
vector, 39
which, 250
which.max, 148
with, 36, 37
within, 147
xtabs, 66
R-squared
    adjusted R^2, 138
R^2, 121
random effects, 264
random effects model, 264
random experiment, 80
randomization, 261
random variable, 81
randomization, 263
reduced model, 141
regression
    pred-res plot, 124
    significance of, 120
    simple linear, 118
regression coefficients, 118
    important, 121
regression model for ANOVA, 169
regression table, 119
regression through the origin, 128
relative risk, 109
repeated measures, 273, 281
replication, 193, 262
rescale, 171
residual, 25, 118
residual sum of squares, 209
round-off error, 171
runs, 291

sampling from a distribution, 98

saturated model, 141
screening experiments, 262
significance, 93
size, 177
skewness, 50
slope, 118
squared multiple correlation coefficient, 121
standardizing, 83
statistical inference, 88
statistical significance, 93
subjective probability, 82
symmetry, 50

test statistic, 89
tips
    adding color, 67
    always compare location and spread on the same scale, 58
    changing plotting symbols and line types, 70
    changing the axis labels, 65
    changing the extent of the axes, 69
    coefficients in logistic regression, 237
    dummy variables for a k level factor, 160
    file name and file path case sensitivity, 31
    hypothesis tests are a self-fulfillin prophecy, 101
    installing R under Windows Vista and Windows 7, 28
    interpreting P-values, 91
    multi-page PDFs and metafiles, 76
    P-values, 90
    quotation marks in R, 39
    retrieving tabular data from web pages, 30
    saving files, 30
    the `par` command, 69
    using R packages, 32

what is so magical about $n = 30$?, 101
treatment, 260
treatment factor, 197, 260
treatments, 260
Tukey's HSD, 174
two-way analysis of variance (ANOVA), 193
two sample t-test, 94
Type I error, 93
Type II error, 177

unbalanced, 276
under-dispersed, 216
unimodal, 17

variable, 8
    covariate, 118
    dependent, 118
    endogenous, 118
    exogenous, 118
    explanatory, 118
    independent, 118
    predictor, 118
    response, 118
variance components model, 264
vector, 74

washed out, 282
weight, 180
weighted least squares, 180
    estimating weights
        residualizing x's, 182
        Welch-James, 180
WLS, 180

Yates's correction for continuity, 108

zero-intercept model, 128
z-scores, 83

Example index

Age and gender of victims of crime, 110
Age estimation from teeth, 142
Annealing of glass, 272

Blood alcohol determination, 278

Calibration in range of fire experiments, 155
Calibration of RI measurements, 153
Carbohydrate deficient transferrin as a predictor of alcohol abuse, 237
Comparing grouped data, 21
Comparing two proportions relating to cranial occipital squamous, 106
Complete separation of the response in logistic regression, 243

Difference in RI between bulk and near-float surface glass, 96
Differences in RI of different glass strata, 94
Differences mean RI in the same window, 170
DNA left on drinking containers, 196, 276
DNA left on drinking containers II, 291
DPD and age estimation, 121
Dummy variables in regression, 163
Dummy variables in regression II, 164

Elemental concentration in beer bottles, 140

GHB concentration in urine, 181
Glass fragments on the ground, 213

Logistic regression for SIDS risks, 236

Manganese and barium, 119
Morphine concentration ratios as a predictor of actuate morphine deaths, 240
Musket shot, 282

Occipital squamous bone widths, 104
Over-dispersed data, 220
Over-dispersion in car crash data, 223

Range of fire estimation, 133
Regression with derived variables, 146
Risk factors for thoracic injuries, 242

SIDS and extramedullary haematopoiesis, 107

Thoracic injuries in car crashes, 223

Using Fisher's exact test, 108

Weighted least squares, 183

For Product Safety Concerns and Information please contact our EU
representative GPSR@taylorandfrancis.com
Taylor & Francis Verlag GmbH, Kaufingerstraße 24, 80331 München, Germany

www.ingramcontent.com/pod-product-compliance
Ingram Content Group UK Ltd.
Pitfield, Milton Keynes, MK11 3LW, UK
UKHW021444080625
459435UK00011B/361